A Lifetime of Moulding Technology and Science Policy in India

A Lifetime of Moulding Technology and Science Policy in India

A Festschrift in Honour of Professor Ashok Parthasarathi

Editor
SACHIN CHATURVEDI

RIS
Research and Information System for Developing Countries
विकासशील देशों की अनुसंधान एवं सूचना प्रणाली

ACADEMIC FOUNDATION

First published in 2017
by

ACADEMIC FOUNDATION
4772-73 / 23 Bharat Ram Road, (23 Ansari Road),
Darya Ganj, New Delhi - 110 002 (India).
Phones : 23245001 / 02 / 03 / 04.
Fax : +91-11-23245005.
E-mail : books@academicfoundation.com
www.academicfoundation.org

in association with
Research and Information System for Developing Countries (RIS)
Zone IV B, 4th Floor, India Habitat Centre
Lodhi Road, New Delhi 110003, India.

Disclaimer:
The findings/views/opinions expressed in this book are solely those of the authors and do not necessarily reflect the views of the publishers.

Cataloging in Publication Data--DK
Courtesy: D.K. Agencies (P) Ltd. <docinfo@dkagencies.com>

A lifetime of moulding technology and science policy in India :
a festschrift in honour of Professor Ashok Parthasarathi /
editor, Sachin Chaturvedi.
pages cm
ISBN 9789332703551

1. Parthasarathi, A. 2. Physicists--India--Biography. 3. Electronics engineers--India--Biography. 4. Policy scientists--India--Biography. 5. Science and state--India. I. Chaturvedi, Sachin, editor. II. Parthasarathi, A., honouree.

QC16.P37L54 2016 DDC 530.092 23

Typeset, printed and bound by The Book Mint, New Delhi.
www.thebookmint.in

CONTENTS

Foreword

This Festschrift is for Ashok Parthasarathi. It is also in many ways about him. This volume has contributions from members of the scientific and engineering community whose own significant contributions to the nation's self-reliance in science & technology—particularly during the prime ministership of Indira Gandhi—have been enabled in no small measure by Ashok, through the role he played as a facilitator–intervener. It is a very well deserved, if overdue, tribute to a man who did not use his illustrious family background, or his positions of influence, for his personal benefit or aggrandisement; but used them for furthering a cause to which he has remained committed, while in office as also when out of it—that of advancing India's self-reliance in science & technology. His contributions to the development of management systems for agricultural research, which will build a collegiate spirit among scientists, are very significant. I have the greatest admiration for his professional and personal qualities.

M.S. Swaminathan F.R.S.
Emeritus Chairman and Chief Mentor,
MS Swaminathan Research Foundation,
Chennai.

Preface

Professor Ashok Parthasarathi has had long association with RIS since its inception. Promoting cooperation in technology development among developing countries has been the major area of our work programme. Our work programme on Science, Technology and Innovation (STI) has benefitted immensely from the guidance and insights of Professor Ashok Parthasarathi.

In the last two decades, debate on S&T policy in India has assumed added importance. The programmes like 'Make in India' and 'Start-up India' have given further impetus to the whole scenario. In this light, this Festschrift, which documents various facets of the important role of Professor Ashok Parthasarathi in promoting India's self-reliance in S&T, becomes a significant contribution. I would also like to take this opportunity to convey thanks to all eminent contributors for enriching the volume by sharing their reminiscences.

Thanks are also due to Academic Foundation and RIS Publication unit for enabling us to bring out this volume so elegantly.

I am sure all those interested in the development of S&T in India will find this volume interesting and useful.

—Sachin Chaturvedi
Director General, RIS

About the Editor/Contributors

Editor

Sachin Chaturvedi is Director General at the Research and Information System for Developing Countries (RIS), a New Delhi-based autonomous think-tank. He was also a Global Justice Fellow at the MacMillan Center for International Affairs at Yale University. He works on issues related to development cooperation policies and South-South cooperation.

Contributors

Sumit Chakravartty, senior journalist and editor of current affairs weekly *Mainstream*, New Delhi. A long standing associate of Professor Parthasarathi.

H.K. Dua, senior journalist, Former Ambassador to Denmark and Member of Parliament. A close friend of Professor Parthasarathi.

Malti Goel, Former Adviser and Scientist 'G', Ministry of Science & Technology and, currently, Chief Executive Officer and Founder, Climate Change Research Institute, New Delhi. She was a student of Professor Parthasarathi when he was Professor of Physics in the Birla Institute of Science & Technology, Pilani, Rajasthan in the mid-1960s.

Killugudi S. Jayaraman, Former Science Editor of the Press Trust of India. A colleague plus associate of Professor Parthasarathi over many years.

C.V. Kalyanasundaram, Former Dean (Academic), Indian Institute of Management and Development, Indiana Group of Institutions, Pune. A co-student of Professor Parthasarathi in the Physics (Hons.) programme of the Department of Physics of Presidency College, University of Madras in the late 1950s.

Rishikesha T. Krishnan, Director and Professor of Strategic Management, Indian Institute of Management, Indore. An associate of Professor Parthasarathi.

B.R. Krishna Kumar, Managing Director, Turbo Tech Precision Engineering Private Ltd., Bangalore. An entrepreneur in hi-tech engineering, whom Professor Parthasarathi had supported in many ways including R&D Funding for developing a unique Low Cost Gas Turbine when Professor Parthasarathi was Special Secretary and Head of the Department of Scientific and Industrial Research, Ministry of Science and Technology in the 1990s.

Chandra Mohan, Chairman, Governing Council, Punjab Technical University (PTU) Nalanda, TQM and Entrepreneurship PTU Mohali Campus, Mohali. He was Founder and CEO for 28 years of Punjab Tractors Ltd., the flagship of the Swaraj Group of Companies. A long standing friend and colleague of Professor Parthasarathi.

Ranjit Nair, Executive Director, Centre for Philosophy and Foundations of Science (CPFS), New Delhi. A close associate-Professor Parthasarathi was a member of the Board of Governors of CPFS for many years.

Roddam Narasimha FRS is DST Year-of-Science Professor, Jawaharlal Nehru Centre for Advanced Scientific Research, Bangalore. He is a Fellow of the Royal Society, and a Foreign Associate of both the US National Academy of Engineering and the US National Academy of Sciences. Member, Space Commission & Director, National Institute of Advanced Study, Bengaluru. A close colleague for over decades.

Geoffrey Oldham, Former Co-Founder of the Science Policy Research Unit (SPRU) at the University of Sussex, UK and Former Director of SPRU for many years. He has been a long standing collaborator of Professor Parthasarathi in the area of S&T Policy Research.

Amrita Patel, Chief Executive Officer, Charutar Arogya Mandal Karamsad, Gujarat. Earlier, she was Chairperson of the National Dairy Development Board (NDDB). In that capacity she worked closely with Professor Parthasarathi on many projects in the application of Electronics to the Dairy Industry.

Prabhat Patnaik, Emeritus Professor, Centre for Economic Studies and Planning, Jawaharlal Nehru University, New Delhi and currently Vice-Chairman of the Planning Board, Government of Kerala. He was a close colleague of Professor Parthasarathi during the latter's stint as Professor and Chairman of the Centre for Studies in Science Policy in JNU over 2000-2005.

A.N. Prasad, Former Director, Bhabha Atomic Research Centre, Mumbai; Member, Atomic Energy Commission, Government of India; Senior Consultant, International Atomic Energy Agency (IAEA), Vienna; and Commissioner, United Nations Monitoring and Verification Commission (UNMOVIC), New York. A close colleague of Professor Parthasarathi in many areas including the Pugwash, Conferences on Science and World Affairs.

Francisco Sagasti, Professor, Graduate School, Universidad del Pacífico, Lima, Perú and Senior Researcher Emeritus, FORO Nacional Internacional, Peru. He worked closely with

Professor Parthasarathi on the International Project on Science & Technology Policy Instruments.

L.K. Sharma, senior journalist and writer. He was the European Correspondent of the *Times of India* based in London for a decade. He reported for five years from Washington as the Foreign Editor of the *Deccan Herald*. He has edited three volumes on innovations in India and has completed a work of creative non-fiction on V.S. Naipaul. A long standing friend and associate of Professor Parthasarathi.

V. Siddhartha, former Emeritus Scientist, Defence Research and Development Organisation, New Delhi. He has been a long standing associate of Professor Parthasarathi.

Rahul Singh, senior journalist and writer. He began his career as the youngest ever Assistant Editor with the *Times of India.* Since then, he has served as Editor of various publications, including the *Reader's Digest*, the *Indian Express,* the *Indian Post*, the *Sunday Observer* and the *Khaleej Times*, Dubai. At present, he is a syndicated columnist and is working on two books. He and Professor Parthasarathi were contemporaries at the University of Cambridge in the UK in the early 1960s and have been close associates ever since.

Krishna Ravi Srinivas is consultant at Research and Information System for Developing Countries (RIS), New Delhi.

George Sudarshan, Padma Vibhushan and Distinguished Professor, Department of Physics, University of Texas, Austin, Texas, USA. A long standing colleague of Professor Parthasarathi.

P. Sudarshan, Currently founder Chairman of Holistic Eye Care Ltd. Formerly founder President and CEO, Technology Development and Information Company of India Ltd (IDICI), nation's first venture capital company and fund. (now known as ICICI Ventures Ltd.) Earlier he was Director Budget and

Economic Analysis of ISRO; Chairman Technology Transfer and Industry Cooperation; Coordinator Propellants, Special Chemicals and Materials and Composites Programmes of ISRO. He was recruited by Professor Parthasarathi in 1970 from IIM, Ahmedabad for the Programme Analysis Group, the planning, analysis and management services Unit of the Atomic Energy Commission.

About Professor Ashok Parthasarathi

Professor Ashok Parthasarathi is a physicist and an electronics engineer by original training and a social scientist and science policy analyst by subsequent training. Having done research in Radio Astronomy at the University of Cambridge in the UK with Nobel Laureate, Martin Ryle, he taught physics at the Birla Institute of Technology and Science, Pilani. He was then a Carnegie fellow at the Massachusetts Institute of Technology, USA where he pioneered research in the then embryonic field of science and technology policy.

Joining the Department of Atomic Energy in 1967 as Special Assistant to the then Chairman, Atomic Energy Commission (AEC), Dr Vikram Sarabhai, he also set up the Programme Analysis Unit, as an inter-disciplinary policy planning and management services unit to the AEC. Concurrently, he worked as a senior staff member in the secretariat of the Electronics Committee of the Government of India chaired by Dr Sarabhai.

In mid-1970s, he joined the Secretariat of the late Prime Minister Indira Gandhi as Special Assistant for Science and Technology at the very early age of 30. For the next five years, and during a subsequent stint in the same capacity over 1980-1984, he dealt with a wide range of science and technology policy, planning and management matters at the highest level. These included the setting up of the Departments of Science and Technology, Electronics and Space, the preparation and implementation of the nation's first comprehensive science & technology (S&T) Plan under the auspices of the National Commission on Science and Technology and several major projects in the areas of atomic energy, space, electronics and defence. He was also involved in aspects of the Bangladesh War of 1971 and the Pokhran-I Nuclear Test of 1974. He is the only person to have held the position of Special Assistant for S&T to Prime Minister, operating from inside the Prime Minister's Secretariat on 24X7 basis.

In 1975, he became Secretary, Electronics Commission and Joint Secretary, Department of Electronics (DOE). Professor Parthasarathi then held the positions of Secretary, Department of Electronics; Secretary, Department of Scientific and Industrial Research (DSIR); Secretary, Ministry of Non-Conventional Energy Sources; and Secretary, Department of Small Scale Industries. During his decade-long tenure in the DOE, he planned and promoted the setting up of 14 professional electronics companies in the public sector, including the nation's microchip company, the Semiconductor Complex Ltd., near Chandigarh.

Professor Parthasarathi has had a long and deep association with the defence sector: whether as a Special Assistant to Dr Sarabhai, or as a Special Assistant for S&T to Prime Minister Indira Gandhi and later as Secretary, Electronics, Government of India (1980-1990). He was closely associated with several major defence projects, such as the Air Defence System, including the Rajendra Radar, which guides the Aakash air-defence missile; the Army Radio Engineering Network (AREN); and several naval systems including the APSOH Sonar; the Very Low Frequency (VLF) Submarine Communication Station in Tamil Nadu; and the 'Advanced Technologies Vehicle' (ATV), the Nuclear Submarine Project. He was also deeply involved in the very conceptualisation and, through many intermediate stages, the final realisation of the ultra hi-tech Special Metals and Super Alloys Plant of the Department of Defence Production, whose import embargoed products are used in rockets and missiles; advanced military aircraft and in numerous areas of advanced electronics. Professor Parthasarathi has similar involvement in the ultra secret area of Electronic Warfare. In all these initiatives, he worked in close collaboration with Defence Science Advisors: Nag Chaudhuri, M.G.K. Menon, Raja Ramanna and V.S. Arunachalam and Chairmen of the Indian Space Research Organisation (ISRO), Satish Dhawan and U.R. Rao.

After retiring from the Government of India in the year 2000, he was invited by the Jawaharlal Nehru University to be a professor in, and Chairman of the Centre for Studies in Science Policy. Over the next five years, he built the Centre as a leading institution in science policy studies in the country.

Professor Parthasarathi has served as Chairman, or member of several government boards and committees, notably, with the governing body of the Consultancy Development Centre of the Department of Scientific and Industrial Research; Scientific Advisory Committee to the Cabinet; and the Planning Commission Task Force set up to draw a plan to make the country a "Knowledge Super Power by 2020".

Professor Parthasarathi is also a member of the Boards of Director of several commercial companies, in both the public and private sectors.

He is well known internationally, having participated in intergovernmental meetings as a delegate of the Government of India, and as a consultant to leading United Nations agencies, such as, the UN Industrial Development Organisation (UNIDO), the UN Conference on Trade and Development (UNCTAD), the UN Environment Programme (UNEP) and the World Intellectual Property Organisation (WIPO).

From as far back as 1970 he has been active in the Pugwash Conferences on Science and World Affairs, launched by Bertrand Russel and Albert Einstein. He has been a member of the International Pugwash Council and its Executive Committee for several years, the first Indian scientist to be in the latter. The Pugwash Movement was awarded the Nobel Peace Prize in 1996.

Professor Parthasarathi was the first Indian member of the international Board of Governors of the South Centre headquartered in Geneva for a decade (1995-2005). The Centre was founded and chaired by the founder President of Tanzania, Mr Julius Nyerere, to advance the interests of the developing countries through both, more coherent interactions and negotiations with the industrialised countries (the North) and greater 'South-South' cooperation.

He is a Fellow of many prestigious international and national professional bodies, such as, the Royal Astronomical Society, London; the Institute of Electrical and Electronic Engineers, New York; the

New York Academy of Sciences; the Third World Academy of Sciences; Trieste, Italy; and the Indian National Academy of Engineering and the Solar Energy Society of India, both in New Delhi.

Professor Parthasarathi has published over 100 papers in prestigious peer-reviewed journals and chapters in books, both national and international. He has written about 25 reports for various United Nations agencies on various aspects of policy, planning and management in relation to science & technology and the environment.

He has delivered a large number of lectures and speeches, both in India and internationally, on a whole range of subjects: science and technology; industrialisation; environment; renewable energy; alternate development strategies for developing countries; and defence and security, including non-military threats to the security of developing countries, etc. He has over 300 publications on these subjects and has edited two books: *Pugwash on Self Reliance* and *Scientific Cooperation for Development: Search for New Directions*. In 2007, he published a unique book on his years as S&T Adviser to the late Prime Minister, Indira Gandhi, *Technology at the Core: Science and Technology with Indira Gandhi.* The book was released by the then President of India, Dr A.P.J. Abdul Kalam, and has sold around 1,700 copies to date, both at home and abroad. Professor Parthasarathi continue to be deeply engaged in each of his areas of interest and expertise.

1

SACHIN CHATURVEDI AND KRISHNA RAVI SRINIVAS

Rethinking Self-Reliance in Science and Technology

Introduction

The role of Professor Ashok Parthasarathi in promoting self-reliance in S&T (science and technology) is well known and through this volume an effort has been made to document it in a meaningful manner. The Science, Technology and Innovation (STI) programme at the RIS has benefitted immensely from his insightful comments and guidance. It is a privilege for RIS to add its tribute to Professor Parthasarathi through this contribution to the Festschrift in his honour.

Until the late 1980s, achieving self-reliance in S&T was coupled with high customs duties, import substitution and achieving a commanding position for the public sector. However, in the last two and half decades, much has changed and self-reliance in sectors like defence, pharmaceuticals and electronics is being talked about again today. Some analysts, however, believe that discussion is couched more in the sense of a pragmatic response to the current international order rather than as an ideological principle. Programmes like 'Make in India' and 'Start-up' are efforts in that direction.

In this contribution, we revisit the idea of self-reliance in S&T as addressed across various contributions in this volume and try to explore the meaning of self-reliance when continuation of the earlier policy framework does not seems to be a possibility, i.e. in the light of the altered circumstances at the national, regional and global levels, the concept of self-reliance in S&T has to be revisited. For instance, in electronics, the increase in imports is largely due to a surge in imports of both consumer and professional electronic goods due to extremely drastic reductions in customs duties, coupled with a seriously erroneous policy decision taken right in 1992-93 that India could be competitive only in computer software but not in computer hardware. From about $28 billion in FY-2011, import of electronics could rise to $40 billion FY-2016.[1]

For some months now, electronics have edged out gold, machinery and pearls to become India's second-largest category of imports, after oil. While the government expects demand for electronics to cross $400 billion by 2020, the industry estimates that at the current rate, domestic production can at best be $100 billion. While some blamed the Information Technology Agreement, to which India became a signatory in 1997, for letting electronic imports flood the country and erode competitiveness, others attributed the rise to an inverted duty structure and the removal of export incentives.

In pharmaceuticals, the dependence on foreign sources of Active Pharmaceutical Ingredients (APIs) puts the sector in a vulnerable position. The bulk drug requirements of the country in 2010 were valued at approximately ₹24,000 crore of which the imports accounted for about ₹11,500 crore.[2] Total import of bulk drugs from China as a percentage of the total bulk drug imports has gone up from 0.3 per cent in 1991 to 47.61 per cent in 2012. In fact, in 2011 it was 57.1 per cent. As far as intermediates are concerned, there is more than 65 to 70 per cent dependence on China, particularly for antibiotics.

The issue is also linked with the future of India's dominance in generic medicines since the formulation industry depends on APIs. Dependence on a single country can affect the industry as that country will be deciding the quantity of supply and price. If that country opts to go for generic production, our companies will find the competition tougher; because of logistics alone they can price us out as the APIs supplied to their local companies can be cheaper and more sustainable.

---

1. The Economic Times, 20 October 2015. *http://economictimes.indiatimes.com/articleshow/49460156.cms?utm_source=contentofinterest&utm_medium=text&utm_campaign=cppst*
2. RIS Policy Brief No. 66, February 2015.

Total imports of APIs and advanced intermediates have grown steadily at a CAGR (compound annual growth rate) of 18 per cent over the last 10 years, from $800 million in 2004 to $3.4 billion in 2014.[3] China alone accounts for 58 per cent of these imports by value. In volume terms, China's share of imports stands at about 80 per cent.

Technological Self-Reliance

In the Indian policy framework, there were three main elements of the concept of self-reliance *viz:* (a) access to technology, (b) institution development and capacity building, and (c) evolving a framework for South-South Cooperation (SSC).

Access to Technology

To achieve self-sufficiency in food, India embarked upon the Green Revolution with the support of the Consultative Group for International Agricultural Research (CGIAR) and other donors. The cooperation between political and techno-scientific establishments stemming from mutual interests was evident in plans for the use of nuclear science and modern S&T in agriculture. In the realm of atomic energy with foreign governments abruptly stopping access to technology and materials after the Pokhran Nuclear Explosion Experiment in May 1974, India relied on its indigenous capacity and capability in materials, technology and skilled personnel to build nuclear reactors and other nuclear systems.

The home grown technological capacity helped India to overcome the import controls imposed by the industrialised countries on our nuclear programme.[4] In space also, since the beginning, the emphasis was on building of home grown capacity and capability in all aspects, resulting in Indian Space Research Organisation (ISRO) developing substantial capacity across all operations related to space. It should be noted that this quest for technological self-reliance was the underpinning principle in S&T policy and practice. Post-1947 the need for institutions and industries to meet the needs from domestic

3. The Economic Times, 29 October 2014. *http://articles.economictimes.indiatimes.com/2014-10-29/news/55559092_1_essential-drugs-bulk-drugs-apis*

4. See Anderson (2010) for details.

production was acutely felt in many sectors ranging from electrical engineering to pharmaceuticals and chemicals. The research system in the public sector in agriculture, medicine and applied research was expanded and new institutions were set up.

Institution Development and Capacity Building

To develop indigenous capacity and to meet national needs without resorting to costly imports, particularly in sectors like pharmaceuticals where the dominance by foreign companies was very high, two public sector companies, Hindustan Antibiotics Ltd. (HAL) and Indian Drugs and Pharmaceuticals Ltd. (IDPL) were set up with foreign technical assistance. For instance, the HAL was set up in 1954, with technology from Glaxo and soon government entered into an agreement with the then USSR for technical assistance and collaboration in producing synthetic drugs, surgical equipment and other pharmaceutical chemicals under a scheme that provided 80 million rubles as loan. Similarly, many public sector units such as Hindustan Machine Tools (HMT), Bharat Heavy Electricals Limited (BHEL) and Bharat Electronics Limited (BEL) were set up and many of them were set up with foreign technical assistance.

The trade relations with the then Council for Mutual Economic Assistance (COMECON) were used to source heavy machinery and capital goods often on the basis of long term loans on concessional terms and denominated in rupees. Thus, India was not solely dependent on what was known as the first world for its needs in foreign technology. Simultaneously, serious efforts were made in developing capacity and fostering cooperation among public sector, private sector and academic institutions. While the Council for Scientific and Industrial Research (CSIR) and other institutions like the Central Drug Research Institute (CDRI) played an important role in helping the fledging generics industry to develop new processes and in reverse engineering, the then University Department of Chemical Technology (UDCT) of Bombay University helped the domestic chemical industry in agro-chemicals and dyestuffs. Although India embarked upon liberalisation in 1991, the thrust on self-reliance was continued.

For example, during the Ninth Five Year Plan, initiatives like, the Programme Aimed at Technological Self Reliance (PATSER) of the Department of Scientific and Industrial Research (DSIR), and the Technology Development Board (TDB) of the Department of Science and Technology (DST) were launched. Techno-entrepreneurship was also promoted. Technology Mission Programmes in sugar technology, bio-fertilisers, bio-pesticides, aquaculture, and fly ash utilisation, etc., were initiated. While the time bound technology development programmes were promoted, a Fund for Improvement of S&T Infrastructure (FIST) was set up as a programme for augmentation and development of infrastructure at universities. For integrating S&T in the socio-economic sectors and promoting S&T for the benefit of society, socio-economic ministries set up 24 S&T Advisory Committees (STACs) in the respective sectors. An Inter-Sector S&T Advisory Committee (IS-STAC) under the DST was set up for facilitating collaboration across sectors and for technology transfer. In specific areas Jai Vigyan Mission Mode S&T projects were launched and these included projects in plant genetic resources, genomics research, ocean thermal energy conversion and new generation vaccines. Although DST and DSIR promoted many programmes and newly established departments like the Department of Biotechnology played an important role in promoting biotechnology, not much happened in technology transfer and development or in successful commercialisation of the technologies developed. While in space, atomic energy and defence project-based funding with accountability worked well, in the case of CSIR, despite much investment and expectations, nothing much had happened in fulfilling the objective of technological self-reliance.

We can learn many lessons from the initiatives in the past and use them to build institutions and for capacity building today. In these times, the role of the State should be that of facilitator rather than one of the agency that directs and controls. As capacity of the private sector has increased considerably in hi-tech sectors, it should be partner in achieving technological self-reliance.

South-South Cooperation

At the global level, the 1970s and 1980s also saw the resurgence of the South as a key actor in the global arena and initiatives like the New International Economic Order (NIEO) sought to redress the global imbalance in economic development. At least some agencies in the United Nations (UN) system were sensitive to the concerns of the developing nations. The Group of 77 (G-77) of Developing countries emerged as a voice that represented the global South. While the Non-Aligned Action Programme for Economic Cooperation among Developing Countries was adopted at the Conference of Non-Aligned Foreign Ministers in 1972, the Buenos Aires Plan of Action (BAPA), adopted at the UN Conference on Technical Cooperation among Developing Countries (TCDC) in 1978, took SSC forward. It spoke of collective self-reliance and called for strengthening existing technological capabilities and creating new ones through the transfer of technology and skills.

In 1981 the Caracas Programme of Action, adopted by the G-77 took this forward and recognised the importance of S&T in SSC. As Professor Ashok Parthasarathi has pointed out that while these initiatives and subsequent programmes launched by UNDP under TCDC and through bi-lateral cooperation resulted in transfer of technology in many instances, these were mostly simple technologies with little science input. He has also raised the question as to whether countries should opt for more of these or more high-technology efforts. TCDC and Economic Cooperation among Developing Countries (ECDC) initiatives had their own limitations but the lessons from these are relevant even today (Parthasarathi, 2006).

Electronics was given importance by the Central government and this resulted in the setting up of many central public sector units like BEL, Electronics Central Electronics Ltd. and the Electronics Corporation of India Ltd. (ECIL). At the state level, some 10 public sector companies in the consumer, professional and defence sectors were established. These companies produced electronics components, consumer electronic goods and catered to the needs of the defence sector. ECIL, BEL and other different initiatives thus enabled creation of the basic electronics sector in India. This was part of broad thrust on technological self-reliance.[5]

As S&T cooperation is pursued in India, Brazil, South Africa (IBSA) trilateral cooperation programme, and BRICS, and with G-77 and China pushing for a Technology Facilitating Mechanism globally, SSC in S&T in these times could make a real difference. How to use this for enhancing technological self-reliance in the South is a topic that deserves an in-depth study.

Challenges Ahead

Our brief review in the above sections indicates that technological self-reliance as a concept was well received in India and was also an important objective in SSC. However, with the ascendancy of neo-liberalism and the process of globalisation and the dramatic shifts in the global trading system with the advent of WTO, policy regimes have moved towards harmonisation and for creating more incentives for intellectual property protection and in a way weakening the options available for the state to intervene. At the same time, the globalisation process resulted in countries like China and South Korea emerging as key players in global trade and innovation.

The technological capacity of the South has increased over the last few decades, despite the gap between developing countries on the one hand, and, between developed and developing countries on the other. Moreover, in terms of R&D spending the gap between the developed countries and developing countries is changing with China emerging as a major spender in R&D, outpacing the spending increase by the United States, and Asian countries emerging as key players in global innovation.[6] In case of India, over the years there have been many official statements pointing out that India can be a global power in S&T but critics have pointed out the problems with India's innovation system and the constraints inherent to the system. The official discourse projects an optimistic picture but critics think otherwise. Leading techno have pointed out that despite several

5. See Parthasarathi (2006) for details.
6. See UNESCO (2015) for an analysis in global R&D spending trends.

successes on the technology front, India relies on imports in more technology intensive products and this includes a range of products from defence equipment to computers and so concludes that successes form just a small fraction of the total technology market in the economy. Further, while the economy has undergone major transformation since the early 1990s, no such change has happened in S&T with the private sector still doing very little R&D and hence another shift in the S&T policy regime is necessary. The merits and shortcomings of the policies of the government in promoting innovation and its intervention in different sectors have been discussed in the literature and comparative analysis with South Korea and China has been made.[7] It has also become clear that in the new settings, self-reliance is the outcome of many factors like capacity of the private sector, favourable external environment, legal and policy changes and opportunities opened up by globalisation.

This volume has enriching contributions from eminent persons from fields of S&T, academia, industry and media, recounting and reflecting on India's journey in the realm of S&T development and the important role of Professor Ashok Parthasarathi in it. They provide significant insights in the evolution of the nation's S&T growth and the tasks that lies ahead.

References

Aggarwal, A. (2001). *Technology Policies and Technological Capabilities in Industry: A Comparative Analysis of India and Korea*. New Delhi: ICRIER.

Anderson, R.S. (2010). *Nucleus and Nation: Scientists, International Networks, and Power in India*. Chicago: Chicago University Press.

Parthasarathi, A. (2006). *South-South Cooperation in Science and Technology for Development*. GLOBELICS Conference 2006. Atlanta.

———. (2007). *Technology to the Core: Science and Technology with Indira Gandhi*. New Delhi: Pearson Longman.

RIS (2015). "India's Growing Dependence on Imports in the Area of Bulk Drugs". *RIS Policy Brief No. 66*. New Delhi.

Singh, L. and K.S. Bhangoo (2013). *The State, Systems of Innovation and Economic Growth: Comparative Perspectives from India and South Korea*. Ludhiana: CDEIS, Punjabi University.

UNESCO (2015). *World Science Report Towards 2030*. Paris.

7. See for example Aggarwal (2001), Singh and Bhangoo (2013).

2

SUMIT CHAKRAVARTTY

Overall Contribution to Science and Technology

Professor Ashok Parthasarathi, who has just turned 75, is not only an eminent scientist but one who has played a pivotal role in harnessing science and technology for advancing India's self-reliance. The rich experiences he gathered while working as Special Assistant to the then Chairman of the Atomic Energy Commission (AEC), Dr Vikram Sarabhai (in 1967-1970) and concurrently as a senior staff member in the secretariat of the Government of India's Electronics Committee, chaired by Dr Sarabhai, were critical to his contributions.

What is most significant, however, is his contribution to the promotion of science and technology in India and their application for the development and security of the nation, in his role as the S&T Advisor in the Secretariat of our late PM, Indira Gandhi, in 1970-1975 and again from 1980-1984. Over both these periods, Parthasarathi handled a wide range of S&T policy, planning and management issues at the highest level. As his CV clearly points out, he was instrumental, with the strong and sustained support of Indiraji, in setting up the Departments of S&T and Electronics and Space; preparing and executing the country's first comprehensive S&T Plan under the aegis of the National Commission on Science and Technology; and several vital projects in atomic energy, space, electronics and defence; besides being directly associated with certain aspects of the 1971 Bangladesh War and the Pokhran-I Nuclear Test of 1974. Till date his contribution in these spheres remains unsurpassed.

This is vividly testified by the observations of another distinguished scientist, Dr Pushpa M. Bhargava, who was the Vice

Chairperson of the Government of India's Knowledge Commission. While reviewing Professor Parthasarathi's book, *Technology at the Core: Science and Technology with Indira Gandhi* (that was released by the then President, A.P.J. Abdul Kalam, in 2007), Dr Bhargava (2007: 23) writes:

> Ashok Parthasarathi's (AP) career has been, in some ways, without precedence. I am not aware of any other case of a scientist being Special Assistant for Science and Technology to a charismatic prime minister of any government in the world (leave aside a democracy as large and complex as India) at the age of 30. It is, of course, rare enough to have such an opportunity to deal with a large variety of subjects (from nuclear energy to food processing), but immensely more difficult to bring professionalism to each area. The fact that AP was able to do this with amazing success, comes through every part of the book. This places AP in a category that would, perhaps, comprise a handful of scientists in the world—if any other at all.

Another feature of Professor Parthasarathi's service to the nation has also been brought into focus by Dr Bhargava (2007: 23-24). In his words:

> There have not been many scientists in the country in the last three to four decades who have put national commitment ahead of personal ambitions, and even staked their career, to ensure that justice is done. AP's rejection of a high level position in the UN Environment Programme (UNEP) at 33, or his being instrumental in the cancellation of the appointment of Thomas Kailath as a consultant to our air defence system, or his passionate and sustained support to indigenous technology (be it super-alloys or tractors) has been exemplary.

Besides his work in the global realm as an active member of the International Pugwash Council and its Executive Committee (the first Indian scientist to be on that committee) for several years, he was also a member of the International Board of Governors of the South Centre in Geneva, founded and chaired by the famous President of Tanzania, Dr Julius Nyerere from 1995 to 2005. Here too, he was the first Indian to be on that Board. The South Centre, was set up with the objective of advancing the developing states' interests through more effective collaboration in their negotiations with the industrialised countries of the North and by fostering enhanced South-South cooperation.

As a scientist of repute, Professor Parthasarathi is a Fellow of several prestigious national and international professional institutions like the Indian National Academy of Engineering and the Solar Energy Society of India (both based in New Delhi); the Royal Astronomical Society, London; the Institute of Electrical and Electronic Engineers, New York; the New York Academy of Sciences; and the Third World Academy of Sciences (based in Trieste, Italy).

Following his retirement from the Government of India in 2000, he accepted an invitation from the Jawaharlal Nehru University to be a Professor at its Centre for Studies in Science Policy (of which he became the chairman). In five years he established the Centre as a leading institution in the country in the area of science policy studies.

At various junctures in his professional life, Professor Parthasarathi has faced concerted onslaughts from people with vested interests, who could not countenance his emphasis on the development of science and technology as a vehicle of the nation's self-reliant progress. These onslaughts were based on misinterpretation of his ideas with the purpose of weakening the nation's advancement in specific areas of science and technology. He weathered all such onslaughts with characteristic unwavering firmness and in the end, emerged unscathed, upholding his integrity without forsaking the fundamental principles that have guided him through thick and thin.

In conclusion, it must be underscored that Professor Parthasarathi has always been an ardent, outspoken and outstanding follower of Jawaharlal Nehru and Indira Gandhi through all his immeasurable contributions to science and technology and their application for the development and security of the nation. While developing science and technology in different spheres, both in the government and outside, he has advanced our self-reliance. I take this opportunity to convey my sincere hope that he will continue his fruitful and selfless work for the nation's progress in the true spirit of Nehru and Indira Gandhi, for many more years to come.

Reference

Bhargava, P.M. (2007). "Unusual Insights into Indigenous Technological Development" (Review of Ashok Parthasarathi's book: *Technology at the Core: Science and Technology with Indira Gandhi*), in *Mainstream*. 7 April. p.23.

3

H.K. DUA

An Architect of Our Science and Technology Policy

In an age when science and technology are leading the world deeper into the 21st century, there are some people in the country who have become unusually vocal. Rather than looking into the future they are looking far back in time and mixing mythology and conjecture with science and technology.

Luckily, there are still a large number of people who are imbued with what Jawaharlal Nehru used to describe as "scientific temper" and who want science and technology to take the country forward. They want India to become a modern nation, have a rational attitude, stand on its own feet and become a power in the 21st century. This cannot be done without using science and technology to tackle our chronic problems—poverty, hunger and unemployment, and all else that goes with poverty. Among such men, whose nationalism is forward-looking and cosmopolitan, and not a narrow-minded exercise in chauvinism, is Ashok Parthasarathi.

Ashok Parthasarathi is a physicist and electronics engineer by original training at Presidency College of the then University of Madras and Cambridge UK, and a social scientist and science and technology policy analyst by subsequent training at MIT, Cambridge, USA. For some four decades, Professor Parthasarathi was passionately involved in the making of our national policy on science and technology. He spent the 1970s and early 1980s as a Special Assistant for Science and Technology to the late Prime Minister Indira Gandhi, operating from Indiraji's Secretariat and making seminal contributions in that position across the whole spectrum of science and technology: agriculture, medicine and environment as well as in atomic energy, space and defence. He also held the position of

Secretary to the Government of India in the following departments: Electronics, Scientific and Industrial Research, Renewable Energy Sources and Small and Medium Scale and Rural Industry.

As a policy maker and administrator in the area of science and technology, Ashok was always keen for India to go for renewable sources of energy, in particular, solar and wind. He gave concrete shape to major policies and programmes in these areas during his tenure as Secretary, Ministry of New and Renewable Energy. He was not impressed by the India-US Nuclear Deal, which he thought promised power but at a high cost and would lead to deep penetration of the sensitive Atomic Energy Programme by a range of US institutions and US policies and practices. But, he did not mind the essence of the deal (particularly in regard to nuclear reactor fuel and nuclear materials more generally) as he felt it liberated India from nuclear apartheid, which had been imposed upon us as a consequence of our peaceful experiment with nuclear explosion in May 1974.

He represented India internationally and has also served as a consultant to several UN agencies like the UN Conference on Trade and Development (UNCTAD), the UN Industrial Development Organisation (UNIDO), the World Intellectual Property Organisation (WIPO) and the UN Environment Programme (UNEP). He was also active in international bodies outside the UN system, such as the International Pugwash conferences on Science and World Affairs, and the South Centre in Geneva. He was keen on the country's, both, purchasing and developing critical technologies so that it could become self-reliant and not depend on the countries in the west, who were always reluctant to sell sensitive technology to India out of fear that India would become a competitor to the western countries. He was deeply committed to an earlier version of the 'Make in India' approach, although the phrase was not in common use then.

Ashok Parthasarathi was responsible for setting up the Centre for Studies in Science Policy in the Jawaharlal Nehru University. He has published 3 books, over 100 peer reviewed papers in scholarly journals, and countless articles in newspapers and magazines and has been passionate in the pursuit of a self-reliant modern India, free from some of the burdens of the past.

That he has a questioning mind was evident in the private dining hall of the India International Centre where a noted institution of the capital, called the Saturday Discussion Group, meets every week to listen to and then discuss with an eminent speaker on a vital issue. He invariably asked the most incisive questions, which others failed to ask, often injecting a new sense of inquiry into the issue under discussion.

The frontiers of science and technology are always expanding, these days at a speed. We need more people like him to take up his cause—the cause of making the world a better place to live in for all its inhabitants—not just for the one billion in the highly industrialised countries but also for the remaining five billion living in the poor countries.

4

MALTI GOEL

My Memorable Physics Teacher, Professor Ashok Parthasarathi

A Quasar is an astronomical entity that emits exceptionally large amounts of radio waves. The name Quasar is a short form for Quasi-Stellar-Radio Source, which existed in the early universe, billions of light years away. In his first Physics class, I remember Professor Parthasarathi telling us about how the discovery of Quasars in 1963 had provided the final piece of evidence in support of the Big Bang theory of the universe. I recall Dr Shiv Yogi Tiwari, Head of the Physics Department at the Birla Institute of Technology and Science, Pilani, walking in with Ashok Parthasarathi to our class of ten students (MSc I Year) in 1965 and introducing him as a physicist doing research in the relatively new field of Quasars in Cambridge, in the United Kingdom. As a seventeen year old, the youngest in my class, I got fascinated by the physics of stars, galaxies and the universe and was motivated to pursue physics more seriously. His teachings made an everlasting impression on me.

I consider it a rare privilege to be able to make this contribution to the Festschrift on Professor Ashok Parthasarathi, my memorable physics teacher. Many years later, in the late 1980s and early 1990s, when Professor Parthasarathi was Additional Secretary and later Special Secretary in the Department of Scientific and Industrial Research (DSIR) I had several occasions to meet him. I was at that time a director in the Department of Science and Technology (DST), and DSIR was on the same campus as DST. Science and technology (S&T) policy paradigms were now his favourite and he had written extensively on science and technology and industrial linkages. Professor Parthasarathi had actively participated in and had a lot of exposure to the deliberations in the United Nations

Conference on Trade and Development (UNCTAD), the UN Industrial Development Organisation (UNIDO) and the World Intellectual Property Organisation (WIPO). His commitment was evident from the tremendous contribution he had made to the growth of public sector electronics industry and exceptionally in setting up of the country's first microelectronics company, called the Semiconductor Complex Ltd. I got an opportunity to seek his insights and guidance in the promotion of science and technology in industry. His analysis of the historical development in science and technology in our country was unique and exciting. I found him very radical in his approach. He encouraged free expression of views, whether it was from students or faculty, on any subject, be it defence, space or nuclear energy. A champion of new technologies, he would never compromise on the nation's priorities. His commitment to self- reliance is strong and deep.

Professor Parthasarathi is my Guru and I admire him. He has published over one hundred scientific papers and several books, of which the most significant is *Technology at the Core: Science and Technology with Indira Gandhi.* This book deals with his decade as a Special Assistant for Science and Technology to our late Prime Minister Indira Gandhi. He presented a copy to me. It is a fascinating primer on policy making in the 1970s and early 1980s, on science and technology in the country. The book provides interesting anecdotes about and the history of decision-making regarding building of capabilities and capacities on science and technology in the country. About the role of universities he has said, "A crucial problem in S&T policy-making, which has dogged us since the 1950s, has been the declining standards of teaching and research in our universities." He also suggested institutional reforms of critical S&T agencies. He used to say, that what we need is "a new vision encompassing a number of key elements and principles, including sustainable development; the preservation of natural and human heritage; 'people-centred' development and grass roots participation; a view of co-operation and a complementary agenda to the new-found global emphasis on competition."

I fondly recall a meeting with him when he was the Secretary, Ministry of New and Renewable Energy (MNRE) from 1995-1998. He emphasised the importance of inter-sectoral planning and coordination for the growth of renewable energy in the country. In DST, after spearheading thrust area projects in physics and having contributed significantly to atmospheric scientific research in the country, I had been given the challenging task of catalysing R&D (research and development) in socio-economic ministries. This scheme had originated when Mr Rajiv Gandhi was the Prime Minister. Professor Parthasarathi was appreciative of the inter-sectoral programme of DST, which I was coordinating. DST had initiated many joint technology projects in partnership with socio-economic ministries and many of these were proving to be success stories in industrial research. Energy efficiency and energy conservation through joint technology development were important objectives of our programme. Professor Parthasarathi remembered me as a bright student at Pilani and believed that there was a vast scope for me to take up the task of promoting renewable energy as well. More than thirty years since he taught us in BITS, Pilani, I found the same excitement, the same zeal to take science and technology forward in the country, and we discussed various options. I value many such inspiring meetings with him.

Energy and climate change are now hot topics for balanced economic and environmental growth. Our country launched an Integrated New Energy Policy in 2006 and its National Action Plan on Climate Change was launched in 2008. Professor Parthasarathi is a strong supporter of renewable energy over nuclear energy as a source of pollution free energy. He was amongst the first to debunk the claim that nuclear energy was a "clean" source of energy. Why? Because almost every stage of the process leading to the generation of nuclear power—from mining uranium ores, to the generation of operating nuclear reactors—produced highly radioactive waste requiring safe re-use or safe disposal. At the recently organised Green Energy Summit, by the India Energy Forum, he fondly recalled working with Prime Minister Indira Gandhi and the insights he gained during that time on the relevance of developing science and technology for the nation. He recollected how Central Electronics Limited was conceived and set up to make crystalline silicon based Solar Photovoltaic (SPV) cells as far back as 1978-1980. Shops called 'Aditya shops' (selling solar energy-based products) were set up by the MNRE to market solar products to the common man. Back in the mid-1990s, in order to optimise land use, a demonstration project for generating 500 KW of energy from both solar and wind sources was set up at the same site. According to Professor Parthasarathi, the current target of developing the capacity to generate 20,000 MW of solar energy and 40,000 MW of wind energy by 2022 was exciting and achievable. Advocating for radical changes needed for improving manufacturing capacity in the country, he suggested that we should make the manufacturing sector more viable, with value additions, and opt for more manufacturing to increase domestic procurement for all our activities rather than depend on imports.

There are several obstacles and challenges that must be overcome to be able to increase our capacity to generate solar energy, including the ability for accurate weather forecasting and availability of smart grids. One hopes that developing science and technology in this area would offer workable solutions. We need scientists who understand the nitty-gritty of the issues; engineers who can translate the new research into practice; and social scientists who can address social transformation and enable peoples' participation issues for achieving the gigantic target for generating solar energy. Translation of fundamental research in solar cells using nanotechnology is still a challenge for industry. Professor Parthasarathi is best placed to advise the nation on how this can be achieved.

5

KILLUGUDI S. JAYARAMAN

Ashok Parthasarathi

From the Diary of a Reporter

It was my good fortune that Ashok Parthasarathi was Special Assistant for Science and Technology (S&T) to the late Prime Minister Mrs Indira Gandhi when I joined the Press Trust of India (PTI) as its first science reporter in 1973. The news agency did not have a full-time science reporter until then.

My job naturally required that I get to know key people in our science establishment and those who influenced the shaping of our national science and technology policies. Parthasarathi, a physicist and electronics engineer, who had studied and researched at Cambridge UK and MIT at Cambridge USA, was the Special Assistant to the Chairman of the Atomic Energy Commission, Vikram Sarabhai before joining the Prime Minister's Office as Special Assistant for S&T to Prime Minister Indira Gandhi, was one such person that my editor Chakravarti Raghavan suggested I should meet first. That is how I came to know Parthasarathi, who remained not only an important contact throughout my three decades of journalistic career in PTI and even after, but also a good friend.

One of my initial stories for PTI was on our first comprehensive multi-sectoral 'approach' developed in 1973 for the S&T Plan 1974-1979, by the National Committee on Science and Technology. Parthasarathi had played a pivotal role in the preparation of this approach. The Plan covered the same period as the Fifth Five Year Development Plan 1974-1979. This was the first time that an S&T Plan for Development and Security had been prepared anywhere in the world. Programmes and projects in the S&T Plan were derived from the socio-economic development plan. However, as regards the policies, the S&T Plan was guided by the approach document prepared in January 1973. My acquaintance with Parthasarathi helped me get exclusive quotes for my story. I remember this distinctly because of the 'dressing down' I later got from Parthasarathi for having quoted him without his permission—a cardinal mistake journalists should remember to avoid. That was a lesson I learnt for which I was thankful. He must have forgotten this incident because soon after, he invited me to cover the 25th Pugwash Conference that he, as the Chairman of the Indian Pugwash Society and member of the International Pugwash Council, was organising in Madras (now Chennai), in January 1976.

As the years went by, my professional interaction with Parthasarathi grew as he moved to hold secretary level positions in the very ministries that I was covering as a reporter for the news agency. These included the Department of Electronics, the Department of Scientific and Industrial Research and the Ministry of Non-Conventional Energy Sources where, I remember, he initiated the steps necessary for India to become a solar power. Discussions with Parthasarathi, a trained physicist and electronics engineer (with specialisation in Radio Astronomy) were always intellectually rewarding. It was natural that on his retirement, Parthasarathi, with a wide range of experience in S&T policy, planning and management, was selected to chair the Centre for Studies in Science Policy at the Jawaharlal Nehru University in New Delhi.

As a reporter, I found in Parthasarathi a quality that reporters treasure most but one that is rare among technocrats at his level. He was frank, honest, and called a spade a spade, even if his views ran counter to the prevailing official policy. For instance, though his science career flowered in the Department of Atomic Energy (DAE), nothing stopped him from being critical of DAE when it came to overall national priorities for the funding of science and technology. He described DAE's ₹30 billion project, Fast Breeder Test Reactor a waste of money and called for "a thorough public debate" on this and other major projects on which DAE was making huge investments without consulting other sections of the scientific or technological communities. He publicly admitted that our S&T priorities were lopsided in favour of the so-called strategic sectors, neglecting R&D in health, biodiversity, meteorology, non-conventional energy and ocean science and technology, areas "that are as important as military security." A clear testimony to his candid nature came after the thermonuclear test, the most important of the two tests conducted on 11 May 1998 and the three small tests conducted on 13 May 1998. Although the DAE claimed that the thermonuclear test (hydrogen bomb) was a great success, Parthasarathi was among those who openly questioned the official claim and declared that test a failure—something very few technocrats would dare to do. (It is a different matter that scientists of the DAE have questioned Parthasarathi's assessment.)

One instance, that stands out in my memory is Parthasarathi's silent endorsement of my premise for an investigative story that PTI released way back in 1974. It was our premise that a US-funded project in New Delhi to control mosquitoes using genetic methods, was actually a cover for developing a system for delivering viruses through mosquitoes in biological warfare. In 1975, Prime Minister Mrs Gandhi ordered to close this controversial project after an uproar in Parliament over foreign funding for covert research. Mrs Gandhi no doubt acted upon the advice of a panel of top scientists whom she had appointed to investigate this project, but my guess at that time was, that as Mrs Gandhi's Special Assistant, Parthasarathi also had a role in the decision taken by his boss.

6

C.V. KALYANASUNDARAM

Reminiscences

A Flashback of My 60 Year Old Association and Friendship with Ashok

Prologue

The other day, my ever-inquisitive 10-year old granddaughter, Diya, asked me about my close friends from when I was a student and about how we spent our time together! This query triggered wonderful memories of the past and, of course, I responded to her query rather briefly, mentioning the several names of which one was, of course, Ashok.

It is indeed a rare privilege and honour for me to dwell on my association with Ashok on the occasion of his 75th birthday.

Loyala College, Madras (now Chennai), 1955—teens from different regions of Madras State (now Tamil Nadu) rather nervously start their college life. Most of us had studied in Tamil medium schools and only a very few had been to English medium schools like Madras Christian College (MCC) High School or Don Bosco.

But, amongst us was a boy who had studied in England before moving to MCC High School in Madras. Yes, it was our own Ashok, a tall and good-looking boy—a lad born with the proverbial silver spoon. No doubt, his background, his physical personality and his studious nature kept us in awe of him (most of us were from middle class families) and many of us were hesitant to get closer to the glamorous boy. As luck would have it, I had an opportunity to initiate a conversation with him during our class on English composition, a subject where both of us scored good percentages! We exchanged information about our 'trainers'—his mother for him, and my father for me. It was somewhat strange that even some of his erstwhile schoolmates were hesitant to be close to him. I guess it was because they were conscious of his background. Nevertheless, I managed to be at least on 'hello' terms with him, till we wound up in the course in 1957.

The scene changes to Presidency College, Madras, heralding a very close association amongst us, stream of Physics Honours! Incidentally, the two Indian Nobel Laureates in Physics—C.V. Raman and his illustrious nephew Dr S. Chandrasekhar—were alumni of the same department!

In later years, the physics department had brilliant students like R. Chidambaram (erstwhile Chairman, Atomic Energy Commission), our Ashok Parthasarathi and K.G. Ramanathan (IAS, retired), who is still active at 75 as an advisor to a major business house.

In 1956 we began as a class of a dozen boys and girls in the stream of Physics Honours. After a year, Ashok went to Jakarta for his summer holidays (his father was our Ambassador to Indonesia!) The second year of the Honours programme saw three boys leaving the course, having gained admission in the Guindi Engineering College. We were now just three boys and six girls in the class.

It all started with Ashok joining me and K.G. Ramanathan for our visits to the nearby Durgha Bhavan or Coimbatore Krishna Iyer restaurants for snacks or lunch and the 'three musketeers', Ashok, Ramanathan and I, started moving together from then on till the final examinations in 1960. Whether it was visits to hotels, movie houses or intercollegiate lectures at MCC Tambaram, we were together in Ashok's car, 'Standard 10', enriching ourselves in terms of academic knowledge and general information.

Looking back, I should say that the kind of close association we had in that impressionable age and critical period in our lives, had a tremendous influence upon me and fine-tuned my tendency to be studious and disciplined. There was no frivolity or unproductive aspect in our discussions, although we did sometimes discuss Alfred Hitchcock movies and cricket, and that perhaps ensured our commendable performance in the final exams. Yes, Ashok, Ramanathan and I were the three of four students who emerged with a glorious First Class out of a class of nine. Unfortunately, we could not celebrate together because Ashok had just flown off to Delhi en route to Cambridge, just a day before the results were announced.

The conclusion of our association during college days notwithstanding, our bond grew stronger in subsequent years, with several opportunities to share thoughts and experiences and well, this still continues.

While Ashok moved to Cambridge, Ramnath was successful in the All India Administrative Services Examination and as for me, I pursued research at the Indian Institute of Science, Bangalore, for my doctorate in physics. Yes, we were in different locations and moving in different directions! Nevertheless, we made conscious efforts to keep our friendship alive.

1960s and the Subsequent Four Decades

The biodata of Ashok Parthasarathi highlights his achievements, areas of expertise and his contributions to development of science and technology. Ashok had all the qualifications and competence to either settle down in a university in the US (MIT) or enter the Foreign Service for a glamorous diplomatic career! But no, he chose to teach physics at the newly formed Birla Institute of Technology and Science (BITS) at a remote location like Pilani, in Rajasthan (Pilani became more easily accessible in later years). What happened after his stint in Pilani is history and Ashok never looked back.

Perhaps, in the history of Indian scientific community, there have been few visionaries and capable administrators and technocrats like Ashok. What has endeared me to Ashok all these years is not the eminence and position he achieved in his career; not his enviable general background; it is beyond mundane issues.

I have always found him simple in outlook, and magnanimous and helpful in attitude. He has never been conscious of status or unduly proud of his parentage (his parents were eminent people but equally simple and magnanimous). Ashok's close friends in his student days, including in his table tennis club, included many boys who came from middle class families and he always made it a point to take us along, whether it be a pilgrimage to Tirupathi with his parents; or to see a first class cricket test match, or to accompany him when he took his foreign tourist-friends on a tour! In essence, he gave us a feeling that we were a part of his family—almost like brothers to him! This was a remarkable personal touch!

Ashok was a national level champion in table tennis. I recall a news item in *The Hindu* announcing his selection as captain of the Cambridge team. He was also a good cricketer!

Ashok is one of those who can speak with clarity and confidence; his insights on international affairs, scientific developments around the globe and his awareness of trends in the management of science and technology in our country were commendable and remarkable! He had a very genuine concern, if not anxiety, for the future of science and technology in the country, even as early as when he was a post-graduate student. I recall receiving a letter from him in the mid-1960s in which he suggested that I talk to my contacts in the scientific community at the Indian Institute of Science and other R&D centres, on this matter and invite suggestions for improvement.

This was his foresight and the later period clearly demonstrated his genuine interest and involvement in the nation's effort in the area of science and technology.

Ashok's competence and capability as an author hardly need any emphasis. There are many instances that present Ashok as a visionary, thinker and objective assessor.

Epilogue

It is my good fortune that I could maintain the link with Ashok for the past 55 years, though we were in different locations. Our meetings were undoubtedly, very rare, but the emotional bond remains strong! It is a proud moment for me, when I respond to queries about him from the younger generation in our family or when friends from various reputed institutes speak highly of him, including from the Bhabha Atomic Research Centre, or the Council of Scientific and Industrial Research and its many laboratories, or the leading computer company in the public sector, CMC, or at any of the 14 State Electronics Corporations also in the public sector such as

Gujarat Communications and Electronics Ltd. (GCEL), or the Kerala State Electronics Development Corporation (KELTRON).

Apart from a group photograph from the farewell at our college, we, the Trio, (Ashok, Ramanathan and Kalyanasundaram) had a very nice photograph taken at the famous Waman Brothers at Madras. I still cherish the memories and have preserved these photographs! This was in 1960; 55 years ago!

7

RISHIKESHA T. KRISHNAN

In Honour of Professor Ashok Parthasarathi

It is a pleasure to write this tribute to Professor Ashok Parthasarathi on his 75th birthday.

While mobility between research, policy-making and practice is common in some western democracies, most notably the USA, it is relatively rare in our country, except perhaps in Economics. But, when one looks back at Professor Parthasarathi's career, a feature that stands out is his ability to straddle these with distinction and be an extremely well-informed and scholarly policy-maker.

The opportunity to work with Prime Minister Indira Gandhi at an early stage, gave Professor Parthasarathi the opportunity to be involved in high-level policy-making at a very young age. The National S&T Plan of the early 1970s, in which he played a major role, was a pioneering effort to link science and technology planning with national development planning. Had this exercise been sustained, our landscape of science and technology may have been very different today.

Professor Parthasarathi's pioneering work on the scientific instrumentation industry shaped his understanding of the challenges in developing science-based industry in India, which he was able to put to good use in his long stint as the secretary of the Electronics Commission and as joint and additional secretary and finally full secretary with the Department of Electronics (DoE).

Professor Parthasarathi's two major stints in managing science and technology policy were at the DoE and the Department of Scientific and Industrial Research (DSIR). At the DoE, he bravely tried to create a strong indigenous capability and capacity in electronics and telecommunication through some 14 state electronics development corporations. For almost a decade, companies such as the Kerala State Electronics Development Corporation (KELTRON) and its counterparts in the states of Maharashtra and Uttar Pradesh, Meltron, and Uptron, the Gujarat Communications and Electronics Ltd. (GCEL) and the Optic Fibre Communication Systems Ltd. (OPTEL), a wholly owned subsidiary of the Madhya Pradesh State Electronics Development Corporation, strove to meet the needs of the country in different product segments of professional electronics and telecommunications. None of their efforts would have been possible without Professor Parthasarathi's strong support through policy interventions, mobilisation of finances from the concerned state governments and the state and central financial institutions, apart from the DoE itself, especially in the area of R&D (research and development), and personal encouragement.

At DSIR, Professor Parthasarathi understood the need to build our capabilities in technological innovation rapidly, in the face of deregulation of the economy. He was a strong proponent of the Programme Aimed at Technological Self-Reliance (PATSER), the first scheme of the Government of India to directly support technological development in industry through financial assistance on the extremely favourable terms of crores of rupees worth financing, per project and with up to 40 per cent of that funding being in the form of a grant. PATSER had a positive influence on a number of subsequent innovation support schemes of the government.

After he retired from the government, Professor Parthasarathi completed the circle by returning to academia at the Centre for Studies in Science Policy (CSSP) at JNU. He expanded the scale and scope of the CSSP, bringing in new people and substantial external funding.

Two strong beliefs have been at the core of Professor Parthasarathi's journey. One is the importance of technological self-reliance for the nation. The second is the primacy of the public sector in not only achieving this self-reliance, but being the core engine of the economy.

While self-reliance may no longer be an explicit goal, the absence of strong technological capabilities in many areas is one of our challenges, even today, as we initiate new programmes like 'Make in

India'. In placing an emphasis on building technological capabilities, Professor Parthasarathi was prescient in identifying an important pre-requisite for the sustainable growth and development of a modern India.

Professor Parthasarathi's belief in the public sector is clearly out of fashion today. But I admire the tenacity and doggedness with which he has continued to stand for his beliefs through his writings in the popular press and involvement in public affairs.

Professor Parthasarathi's career clearly shows us how mobility between research and policy can enrich policy-making and public discourse. It would be a fitting tribute to his contributions to S&T policy if we could find more ways to formally enhance such crossing of boundaries.

8

B.R. KRISHNA KUMAR

Ashok Parthasarathi

A Great Scientist, Philosopher and Guide

Professor Ashok Parthasarathi and I have had a working relationship spanning more than two decades. He is a man of many parts. To me, he has been at different times a senior government official, a stern disciplinarian, a thought-provoking author, a genial mentor, an insightful director on the Board of Directors of Turbo Tech and a fascinating friend. His razor-sharp intellect cuts to the core of any problem and his swift and sure insights and advice are always valuable. His repertoire of engrossing personal anecdotes, involving the movers and shakers of the polity, makes him a most welcome guest in any gathering. His is one of the relationships that I value most in my professional and private life.

I first met Professor Parthasarathi in 1995 when he was Additional Secretary, Department of Scientific and Industrial Research (DSIR). My company was in the process of developing the nation's first industrial gas turbine, and we were looking for financial support for our project. We approached several banks, financial institutions and venture capital companies, and were uniformly and brusquely turned down. There was no interest in funding technology development by a small local company.

At last, I approached P. Sudarsan, the head of TDICI (the venture capital arm of ICICI Bank), a good friend and mentor. He instantly suggested that I speak with Ashok Parthasarathi, the head of DSIR. Sudarsan had been recruited by Professor Parthasarathi to the Department of Atomic Energy as a fresh graduate from IIM-Ahmedabad, and they had remained good friends ever since. He told me that if there was one steadfast champion of indigenous technology development in government, it was Professor Parthasarathi. He offered to write a letter of introduction, an offer I gratefully accepted.

A week later, with great trepidation, I called Professor Parthasarathi at his office. Frankly, I had no great hopes, as my experience with government officials up to that point had convinced me that 'those who can, do, and those who cannot, take refuge in government'. I dreaded the task of explaining to a technologically ignorant 'aparatchik' the intricacies of gas turbine technology and my project. The call was an eye-opener. The person I spoke to was obviously a well-read scientist who grasped within a few sentences the essence of my project. He was aware of the strategic importance of gas turbine technology and the technical challenges involved in developing a gas turbine. He agreed with me that the time had come for the nation to get into gas turbine technology. Professor Parthasarathi had received the letter from Sudarsan, and trusted his judgement of me and my company. He invited me to Delhi to meet the DSIR team, and asked me to apply for support under the Programme Aimed at Technological Self-Reliance (PATSER). Of course, I did so speedily.

Steering the application through the maze of government reviews and approvals was something I had no experience with. However, with Professor Parthasarathi's bold, strong and sustained support, I was able to navigate the shoals of government without incident. The sharp and sometimes scathing criticism of my submissions by Professor Parthasarathi made me hone my plans and thinking to a razor's edge. His incisive questions removed all the fuzzy thinking and vagueness from my plans. Those interactions were sometimes a matter of fear and dread, as he certainly did not suffer this fool gladly. At the same time, I realised that this was a trial by fire, and I would emerge the better for it. Fundamental to the interaction was mutual trust, the fact that he was attacking issues not personalities, and that he was on the 'same side of the table', as a young technologist.

The presentation of our proposal to the DSIR screening committee is etched in my mind, as that was one occasion when Professor Parthasarathi sat on the other side of the table, and I was on my own. Having already answered all the hard questions posed by Professor

Parthasarathi, responding to the questions posed by the committee was a walk in the park. The committee approved the project and funding was granted, no doubt owing to Professor Parthasarathi's efforts. In those days (the early 1990s), for someone in the government to actively support a small private company would lay him open to misunderstanding and attack by ill wishers in the hide-bound bureaucracy. What he did, required courage and a firm belief in the cause of indigenous technology development. In the event, the gas turbine development project has been a story of successfully commercialising a domestically-developed advanced technology.

Subsequently, after his retirement, I invited Professor Parthasarathi onto my Board of Directors. His vast network in the science and technology community has opened many doors and helped the company through the years. The success of my firm in working with Defence Research and Development Organisation (DRDO) is owed in large part to the valuable advice we received from Professor Parthasarathi.

One is privileged to meet only very few truly outstanding human beings in one's life. For me, my relationship with Professor Parthasarathi is something I will always treasure.

9

CHANDRA MOHAN

Four Decades of Association with Ashok Parthasarathi

Commitment to the nation among students of my generation, who were in college during the Partition, was unbelievable. And, Pandit Nehru with his red rose was the unquestioned idol. Cycling 30 km to hear him speak on a sizzling afternoon in May was a routine. A few of us travelled all the way to Delhi by overnight train to see him unfurl the flag on the ramparts of the Red Fort on the first Independence Day, 15 August 1947.

Pandit Nehru's knowledge of science and technology and its role in the strategic, economic and social development of a country was unequalled among politicians of that generation. It was that conviction of his which led us to invite the best of brains in the world and give them total freedom to help us build giant dams like Bhakra, educational institutions like IITs and IIMs, chains of research and development (R&D) labs in agriculture, medicine, industry and atomic energy and special financial institutions to promote industrialisation. His romance with science and technology also left a deep imprint on his young daughter, Indira.

When Mrs Gandhi became the prime minister in 1966, she renewed Nehru's thrust on science and technology with greater vigour. Her first major step was inducting young science-minded ministers in key portfolios: C. Subramaniam, the father of the Green Revolution, was first inducted in the ministries on planning and science and technology, and subsequently in key ministries on industry, finance and agriculture; Mohan Kumaramangalam in steel; and R. Venkatraman in planning. This was followed by induction of top-notch technocrats in key positions: Mantosh Sondhi in steel, Lovraj Kumar in petroleum, M.S. Swaminathan in agriculture, and M.G.K. Menon in electronics. The results of this thrust were astounding:

- Green Revolution enabled a famine-ridden beggar nation to lift its head with confidence and pride for the first time. It has never looked back on the food front ever since.
- National Dairy Development Board led to the milk revolution.
- Swaraj and Punjab Tractors Ltd. (PTL)'s completely indigenous Indian tractor made with 100 per cent Indian technology succeeding against the best of global brand names. But for the vision of and nurturing by C. Subramanium and Mantosh Sondhi, against the advice of the civil services, it would have been strangled at birth.

Another interesting initiative of Mrs Gandhi in the mid-1970s was the induction of young Ashok Parthasarathi, a scientist in Radio Astronomy from Cambridge and MIT, as her special assistant on S&T. This led to the setting up of the National Committee on Science and Technology composed of national brains, assigned with the task to draw up a national policy document on S&T called the Approach to the S&T Plan and a Comprehensive Multi-sectoral Plan based on the approach for policy and substantive programmes and projects in each of 24 sectors of the economy. The plan led to:

- establishment of autonomous structures for S&T institutions inclusive of finance, recruitment and procurement to keep them away from bureaucratic meddling;
- establishment of departments of science and technology, electronics and space;
- recognition of electronics as the key technology of the future and establishment of an Electronics Commission to oversee promotion of industry in advanced electronics; and
- setting up of a semi-conductor complex in Mohali, near Chandigarh for manufacturing micro-chips.

I got to know Ashok in 1975 when, despite the wide acceptance of the 100 per cent Indian Swaraj tractor by our farmers, PTL was struggling for financial survival because of unfair price advantage to

Completely Knocked Down Kits (CKD) import-based competitors. It was his unearthing of a decision by the cabinet, clearing every form of financial support for commercialisation of indigenous technology and, behind the scene work with the Finance Minister, C. Subramaniam, which led to the unprecedented decision of waiving 10 per cent excise duty for Swaraj for two years, against the advice of his secretaries. Swaraj never looked back and since then grew into a nationally and internationally known blue-chip company.

Ashok's nose for advanced technologies and passion for putting the country on the global map is unbelievable. His elevation to higher levels in assignments as joint, additional and full secretary led to national and international involvement in advanced technology and its induction into the country in every domain.

Both of us pooled our strengths to undertake a mission to propel national technological capability forward, whenever needed.

Organising the M.M. Suri Memorial Lecture in Chandigarh, in 1991, was one such occasion. The memorial lecture delivered by Dr Kurien of the National Dairy Development Board (NDDB), the vehicle of the White Revolution, was presided over by Minister C. Subramaniam and compered by Ashok. Listening to the personal experiences in leading major revolutions in technology, by three eminent scientists on one platform, was a great education.

Joining of our forces over the prevailing national norm of licensing field-proven technologies took a quantum technological jump in 1996. It was for commercial up-scale of technology.

This world-class technology with 20-global patents had been developed after 10 years of effort by a sibling of the world-famous Bell Telephone Laboratories, USA. Because of its overwhelming superiority for mobile applications, global interest in it was large and licensing fees steep. Lump sum down payment was to the tune of $1.5 million; royalties cost 5 per cent. Such licencing terms were unheard of in India.

With no clear-cut major user yet, the cost of a versatile Pilot Plant for covering a wide range of applications, including environment-friendly, battery-operated vehicles, came to about ₹300 million.

Despite my PTL credentials with the financial world, and Ashok's with the electronics industry, no private partner was willing to invest. Even GoI's own Technology Development Board together with national financial institutions found the risk too large. It took all of four years for both of us to get them together to evolve a financial structure acceptable to all.

Setting up in Mohali only commenced in 2000, with me serving as the CEO and Ashok serving as a key member of its high-power national board.

A Pilot Plant for a new atomic-level, mass-production, and high-precision technology, on a shoestring budget meant de novo design, local development and process-proving of an entire range of hi-tech equipment. But the team did it. Tooling-up for each size being expensive, batteries for a limited range of cell-phone models were launched commercially in December 2004. Brand-named, they also received excellent commercial reception for their extended talk-times.

But tightening of purse-strings by the financial institutions left me no option but to resign from the board in October 2005. Institutions shut down the plant soon after. But, despite the fact that it was a national endeavour, legal notices from the financial institutions dogged us for years.

Loss of over 10-years of colossal effort hurt us both. More so, when we see the same Lithium Ion, polymer battery technology, powering every cell-phone and iPad in the world today. Elon Musk's battery-operated cars also use the same technology.

Our experience establishes that the courage to invest resources to become equal partners in the global race for technology is not an easy game for any nation.

We both, on our part, have forgotten that unhappy chapter. And undeterred, we continue to pursue our mission of raising national technological capability in our own separate ways. Ashok, through papers, books and lectures, on a wide range of his national and international experiences; and I through propelling world-class quality and entrepreneurship.

10

RANJIT NAIR

Technology to the Fore

Institution Building by Professor Ashok Parthasarathi in a Post-Colonial India

The creation of world-class institutions is an aspiration that entertained us for over a century. It acquires certain poignancy when an Indian studies abroad at an iconic western centre of learning. It is not unusual for many students of the latter, including me and some of my close friends and fellow alumni, to indulge in some soul-searching about the absence of such environments here in India and visualise various possible scenarios. Some of us wondered about our teaching institutions of yore presently only existing *nāmamātra*, in history books, and whose memories have been all but lost. The names of Takshashila, Nalanda, Vikramashila, Valabhi, Pushpagiri, Odantapuri, Somapura, Kanchi, Mithila and Ujjain in the north would be routinely invoked as would Amravati, Kanchipuram, Madurai, Matilakam and Kandalur in the south, even though few, if any of us, had visited any ancient ruins as might exist of these redoubts of higher learning, which are at some distance from the metropolises of modern India.

A harmless enough pastime, considering that one could have done infinitely worse. Choudhary Rahmat Ali, a Cambridge-educated barrister who dreamt up the witches brew of Pakistan, was rudely rebuffed by Muslim thinkers in India under the British rule. The latter regarded the idea as impractical and a calculated move that painted them in communal-revanchist colours, in order to oust them from mainstream politics. Jinnah even claimed that the 'P word', which some silly students in England had coined, was foisted on them by the Hindus and the British. In hindsight, we know that these critics were right and that Muslim citizens of colonial times and later, of the successor states, have suffered, along with the Hindus who were portrayed as the principal enemy.

The idea of an institution of learning that would command respect worldwide is nowhere as likely to raise hackles—but only just. The centre I helped found, decided to document the learning from experiences of second generation of institution-builders in a series entitled, *The Republic of Science*. The series was launched to mark the golden jubilee of our Republic, at the India Habitat Centre (Nair, 2007). As the series unfolded, I got to know Ashok Parthasarathi, who evinced keen interest in it. He was then making a return to academia as the head of the Centre for Studies in Science Policy at Jawaharlal Nehru University after retiring from a distinguished career in the government.

"A policy is a temporary creed liable to be changed, but while it holds good it has got to be pursued with apostolic zeal." This dictum of Mahatma Gandhi may aptly be applied to Ashok, who has zealously devoted himself to the cause of Indian science and technology ever since he returned to India in 1967 from MIT, to work with one of the stalwarts of our science and technology establishment, Vikram Sarabhai (Parthasarathi, 2007). After Homi Bhabha perished in a mysterious plane crash in the Alps in 1966, which a former CIA operative Robert Crowley avers was an assassination, Sarabhai took charge of the Atomic Energy Commission (AEC) as its chairman, in addition to his responsibilities for setting up the space programme.

The scion of a business family, Sarabhai was one-up on Bhabha in a vital aspect of institution building. Bhabha had to importune his uncle, who headed the Dorabji Tata Trust, for funds to establish the Tata Institute for Fundamental Research in 1945, which went on to become the nursery of the atomic energy programme. In contrast, Sarabhai could set up the Physical Research Laboratory in his hometown in Ahmedabad, from his family's own resources. This laboratory went onto play the corresponding role for the space programme, which he launched in 1962. Ashok had met Vikram Sarabhai at MIT whilst on a Carnegie Fellowship in 1966-67 and they discussed about setting up a Programme Analysis Group in the

Department of Atomic Energy. Sarabhai was positive, as he had also inherited (from Bhabha) the Electronics Committee, which was tasked with implementing recommendations of the Bhabha Committee on Electronics. He thought that Ashok's training in electronics engineering and radio astronomy at Cambridge, UK, would come in handy in this area. Ashok came in as a fresher without any baggage and could examine, with some objectivity, the developmental plans of the AEC. Ashok found Sarabhai's 10-year perspective overambitious and unachievable, and this led him to differ from his mentor. Sarabhai could have taken a leaf out of Winston Churchill's book. The latter quipped, "I always avoid prophesying beforehand, because it is a much better policy to prophesy after the event has already taken place."

The scientist is not temperamentally suited to negotiate the corridors of power as a civil servant. He faces more challenges than a generalist recruited on the basis of the Union Public Services examination. On the one hand, a lateral entrant is often bitterly resented by professional civil servants as an unwelcome interloper. On the other, professional scientists could thumb their noses at a person they regard as a *parvenu*. The first generation of institution-builders enjoyed direct access to Prime Minister Jawaharlal Nehru, who had taken the Natural Sciences tripos at Cambridge University, and was a firm believer in science and technology as the basis of both country-wide industrial development and modern rational thought that would serve as an antidote to traditional superstitions. Sheltering of scientific agencies under the PM made decision-making a whole lot easier and quicker. Nehru's daughter Indira Gandhi, who was sworn in as PM after Nehru's successor Lal Bahadur Shastri passed away in Tashkent in 1966, did not pursue higher education but she took to heart the Nehruvian vision of science and technology. When she came to know Ashok, who was active in arranging conferences with leading scientists as part of his work on programme analysis, she offered him a job as her special assistant on science and technology. This post was created for the very first time and at the age of 30, Ashok found himself at the core of not only the S&T establishment but also the government as a whole. During

Indira Gandhi's halcyon years, she took a great deal of personal interest in building national capabilities across a variety of sectors, which were only dimly presaged in Nehru's time. From electronics to the environment, she took decisions with expert advice, with the alternatives set out very clearly to her by Ashok, who apprised her of the developments in key areas. The feuding chiefs of the scientific establishment were often nonplussed when, after a discussion, Indira Gandhi would turn to Ashok and seek his opinion. She would then end the meeting saying "let's go along with what Ashok says."

At a young age Ashok had the opportunity to serve the cause of S&T development from the centre of political power, as a policy adviser and analyst who was fundamentally committed to the cause of indigenous development. With big-ticket projects, large sums of money are involved and it is important for the country to ensure transfer of technology as an essential part of the process. The acquisition of indigenous capabilities was an absolute must, lest the country be held to ransom by suppliers. An uncritical adulation of the private sector, and a corresponding denigration of the public sector in poor countries, was part of the unquestionable wisdom that international lenders such as the IMF dinned into poor countries that ardently wished to be in their debt. The powerful played by different rules; they could do what they liked, but vassal states had no business to ask uncomfortable questions. The nation's progress, now much-lauded, arose from an uncompromising insistence that the sovereign autonomy of the country had to be respected. Quite often, projects incurred cost and time overruns and had to be nursed back to health, as is the practice followed universally. Every major demand on the exchequer had to be carefully addressed and adjudged without fear or favour. From the heights of Raisina Hill, Ashok would monitor and promote the country's S&T system across the board, intervening as and when necessary. *Ars longa, vita brevis,* but things just do not happen in a hurry. Patience is needed in the development of science and technology, because the gestation period, even for the development of a prototype, is likely to be long. Unless funds are provided to take a device from the drawing board to the real world, public sector undertakings will appear like errant schoolboys who know they have blotted their copybook, especially when there are cost and time overruns. To be a responsible public servant is every bit as difficult as it is to be a critic from the outside. Not everybody has the license of a writer. George Bernard Shaw once said, "I make a fortune from criticizing the policy of the government, and then hand it over to the government in taxes to keep it going".

Jawaharlal Nehru favoured bold moves when he said, "The policy of being too cautious is the greatest risk of all." This is a handy maxim when it comes to fulfilling the historic aspirations of the people of our vast and diverse land. In a world that is incessantly transformed by technology, we have to run to be in the same place, entering as and when necessary.

References

Nair, Ranjit (ed.) (2007). *The Republic of Science: Golden Jubilee Vignettes*. New Delhi: Scientia.

Parthasarathi, Ashok (2007). *Technology at the Core: Science and Technology with Indira Gandhi*. New Delhi: Pearson Longman.

11

RODDAM NARASIMHA

In Honour of Ashok Parthasarathi

It is a great pleasure and privilege to write these few words on the occasion of a Festschrift in honour of the distinguished scientist and science policy analyst, Ashok Parthasarathi. I have had the good fortune of knowing him for some three decades or so, and of having many discussions with him on a variety of science policy issues of common interest to both of us, including in particular, those related to aeronautics and space. Ashok has been at the centre of many important debates that have taken place within the government, as well as in public, and his views have greatly influenced decisions made by the Government of India on issues connected with science and technology, especially when Mrs Indira Gandhi was the prime minister and he served as her Special Assistant for Science and Technology. Ashok's analyses have always been sharp and clear and have been driven by the needs of the nation. His views have been heard with respect, not only by leaders in India, but abroad as well, in particular by many international agencies. In later years, he contributed to the management and growth of the Centre for Studies in Science Policy at the Jawaharlal Nehru University. His abilities as a physicist and his expertise in electronics enabled him to understand a wide variety of modern technologies. His later specialisation in science and technology policy analysis helped him look at complex problems with clarity, and with insight into the special needs of a developing country that is trying to achieve self-reliance and follow an independent foreign policy in a world dominated by the Cold War.

From the many works that have come from his pen, the book *Technology at the Core: Science and Technology with Indira Gandhi*, published in 2007, stands out as an account of science policy making in the country, during Mrs Gandhi's tenure as prime minister.

Ashok was in a particularly strong position to write this book because of his official role in the Prime Minister's Office and involvement in many policy decisions that were made at the time. His account is an invaluable source of material for both, experts as well as the layman interested in the dynamics of policy making in India. Furthermore, the period during which Ashok served the prime minister, namely 1970-75, was a particularly eventful one, what with the Bangladesh war in 1971; establishment of the Department of Space in 1972; and a nuclear test in 1974, all of which had great consequences for the affairs of the Indian subcontinent, and indeed for the rest of the world. Ashok's account of these and other events during that period is illuminating, insightful and fascinating. One thing that stands out is how, among all the players involved, Mrs Gandhi was one person who always had a long-term view, an appreciation of the big picture, and a clear vision of where the country ought to go. After her father, she was the one prime minister who took almost a personal interest in the distinguished scientists who served her government. I have heard it said, that on occasions when there was a difficult issue regarding science with her colleagues and assistants, she would say something like "please do not play around with my scientists".

The book is full of fascinating stories about factors that went into the making of policy decisions, in particular, in the atomic energy, space and defence sectors—all of which have great strategic importance. Ashok traces how both principles and personalities play their role in influencing decisions. Homi Bhabha, Vikram Sarabhai, P.N. Haksar and Satish Dhawan were all, in one way or another, participants in decisions about the future of the nuclear power programme; development of electronics; the kind of reforms needed concerning the scientific establishments; the possible relations between the Defence Research and Development Organisation (DRDO) and Indian Space Research Organisation (ISRO)—given Dhawan's firm conviction that the two should be separate; and so on. A particularly interesting side line is about the way Dhawan succeeded Vikram Sarabhai as head of the space programme after Sarabhai died at 52. When he was invited to head the space programme, Dhawan wrote in his reply, an outline of what we should be doing in the space programme and how it should not have any connection with the DRDO. Effectively, he said that he would be honoured to head the programme only if the government accepted his prescriptions for it. There is also a discussion of a very interesting proposal made by Sarabhai to the Administrative Reforms Commission in the late 1960s. While most agency heads seemed satisfied with the prevailing system, Sarabhai suggested a ministry of advanced technologies that would be in charge of atomic energy, space, electronics, aeronautics and earth sciences (including the oceans), each with a commission of its own along the lines of the one which had been set up for atomic energy. Ashok points out that the proposal had no takers and that "it died instantly". But in actual fact, it survived and later on led to the establishment of the Electronics Commission in 1971 and the Space Commission in 1972. The Earth Sciences Commission came 35 years later, but has not had the same distinguished history as its predecessors in atomic energy and space. For almost half a century, there has been constant policy discussion about setting up of the Indian Aeronautics Commission. However, till date no concrete action has been taken in this regard nor any alternative has been suggested.

For anybody interested in Indian science and technology policy, Ashok Parthasarathi's book is a compulsory reading.

There have also been many articles that Ashok has written on specific issues, in various journals here and abroad. Let me take only one of them, which has been of particular interest to me, namely one on the Light Combat Aircraft (LCA) project. The project has seen many ups and downs, has received a lot of publicity for so-called 'time and cost over-runs' and contradictory announcements by successive national governments about how close it is to final operational clearance. This is not the place to discuss the LCA issue in detail, but I should draw attention to an account which appeared in *The Hindu* on 9 March 2008 (updated 10 February 2011), co-authored by Ashok with Vice Admiral Raman Puri (who retired as the Chief of the Integrated Defence Staff in Delhi). Anybody who wants an appreciation of this long drawn out project must read this article. From what I know through my own involvement in the project, the analysis of Parthasarathi and Puri is a compulsory reading. If you see another piece in the media on the subject, I would strongly recommend that you should open up to what Parthasarathi and Puri say in that article. I only quote here, one of their conclusions:

> As for development costs, the LCA has remained still within the sanction of $1.2 billion—which is about the lowest anywhere. As for the time over-run in the strict sense it is only by a year or two despite the [US] sanctions [following the Pokhran-II test] . . . a first-of-type aircraft of this degree of complexity has not been developed anywhere in the West or in Russia in less than two to three decades.

Rocket science is popularly thought of as something extremely difficult and daring, but is it not paradoxical that the Indian rocket programmes, in space and increasingly in defence, have been generally successful while the combat aircraft programmes generally not? It is almost as if the Indian aerospace system was declaring, "If it is not rocket science we cannot do it"!

Let me conclude by saying that I cannot think of any other scientist or policy analyst in India, who has had the unparalleled experience in the corridors of power as has Ashok Parthasarathi. He has had the opportunity to observe how all those momentous decisions were made, what influences drove those decisions, and knows the background to such decisions. His account is sometimes elevating, sometimes depressing but always riveting. I hope to hear more from him on science and technology policy in India in the coming years, and I wish him many happy and satisfying years of analysis and argument in the service of the nation.

12

GEOFFREY OLDHAM

A Tribute to Ashok Parthasarathi

Both Ashok and I trained as natural scientists. Ashok was a radio astronomer applying the methods of physics to understand outer space. I was a geophysicist applying the methods of physics to understand the interior of the earth. But for the past 50 years, we have both focussed on how science and technology can be mobilised to help solve societies' problems on the surface of the earth.

During these 50 years, our paths have crossed frequently and we have become good friends. I have learned a lot from Ashok about how science and technology have contributed to the development of India. A particular pleasure has been the interaction between our families.

In this book, there will be much that is written about Ashok's contribution to science policy in India. I, however, will focus primarily on Ashok's international contributions.

Indian Scientific Instruments Project

Shortly after the establishment of the Science Policy Research Unit (SPRU) at the University of Sussex in the UK, we received a visit from a young Indian radio astronomer who had become interested in science policy. This was our first meeting with Ashok Parthasarathi, who was on his way back to India for a position as Special Assistant to Vikram Sarabhai, Chairman of the Atomic Energy Commission in Mumbai. It was an interesting and lively meeting.

One of the issues we discussed was how science based industries in developing countries acquired the technology needed to manufacture. Specifically, we discussed how the scientific instruments industry in India might have acquired its technology. We identified several possible routes. For example, the technology might have originated in India through government research establishments, or by research conducted in the private sector. Alternatively, it might have been developed abroad and transferred to India by different routes. For example, the foreign company who owned the technology might have established a subsidiary in India, or it might have established a joint venture with an Indian company, or the technology might have been transferred through a licensing agreement. Finally, the technology might have been pirated and developed in India by reverse engineering.

There were pros and cons for each of these routes but Ashok confirmed, that as far as he knew, there had not been a detailed survey of what had actually happened in the Indian scientific instruments industry. We continued our discussions on this issue after Ashok returned to India and eventually agreed to collaborate in a joint project aimed at identifying what had happened with the scientific instruments industry in India and what had been the pros and cons of the different routes for acquiring technology.

It was the first time that SPRU had collaborated with researchers in a developing country and the collaboration process was not always easy. We were determined that this project would not be like many other collaborative projects where the foreign team designed the project and the local team collected data, to be analysed by the foreign team and also published by them. This project had to be a genuinely collaborative effort. At times the SPRU team in India felt that in trying to have genuine collaboration, the Indian team had gone too far in trying to control the activities of the SPRU team. Eventually, I met with Ashok and we worked out a modus operandi that was satisfactory for both teams. This was not the last time when Ashok and I negotiated over what at the time were considered contentious issues. But we always found solutions!

The Indian Scientific Instruments project was a pioneering science and technology policy project that helped us understand the pros and cons of the different routes for Indian companies to acquire foreign technology. It contributed a great deal to SPRU's understanding of a major aspect of how science and technology contributed to development. Through Ashok, it also helped influence Indian policy regarding this topic.

SPRU's Tenth Anniversary Report

When SPRU reached its tenth anniversary, the University of Sussex set up an external ten-year review committee. Its report was intended to strongly influence the University's decision on whether to continue the Unit's existence. Members of the committee were experts in each of the themes that SPRU had been doing research on. The University asked each member of the review team to assess the quality of SPRU's research in their area of expertise as well as its impact on policy. Ashok was invited to review SPRUs work on development, and Lord Flowers, the rector of Imperial College, was invited to chair the committee.

Ashok's comments on the Developing Country Programme were very positive regarding both the academic quality of the research and on its impact on policy. He was able to draw on his experiences in India and his knowledge of the international impact of the work of SPRU on United Nations Conference on Trade and Development (UNCTAD), the Organisation of American States (OAS) and United Nations Development Programme (UNDP). He was also aware of our impact on Canadian science policy.

His review was in contrast to the reviewer of the Unit's research on industrial innovation. He gave high marks to the quality of the research, but low marks for its impact on UK policy. This low impact on policy was ascribed to the fact that at that time (mid-1970s) much of the data needed to comment on and influence policy was confidential and not accessible to SPRU.

The Flowers report was to have a considerable impact on the future of SPRU. It had recommended that SPRU be continued, but only if the University provided more financial support to the Unit. Several years later the chairman of the Social Science Research Council (SSRC) challenged the then vice chancellor to implement the recommendations of the Flowers report. If he did not, then he suggested that it would be difficult for the SSRC to continue funding the Unit.

This had the desired effect, and the University did implement the recommendations and started to provide more resources to the Unit. We clearly owe special thanks to Ashok for his part in ensuring the survival of SPRU.

The Sussex Manifesto

In 1970, a team from SPRU and the Institute of Development Studies at Sussex University were asked by the UNDP to prepare a report summarising the current understanding of the policy issues concerning the application of science and technology for the benefit of the developing world. The UN published this report as a discussion document.

At about the same time, Ashok and his colleagues were preparing the Government of India's Approach Document on science and development in India. This was to provide the basis for the preparation of a new policy on science and technology. The Approach Document explained why it was appropriate for India to become technologically self-reliant. This position was strongly opposed by some Indian economists.

The fact that the UN had published the Sussex report (known in some quarters as the Sussex Manifesto), which itself had made the case for technological self-reliance, provided useful support to the Parthasarathi team. It is interesting to note that the Indian economists who opposed Ashok's position tried to sabotage the science and technology approach document by saying that it was based on the findings of a UK report! Nothing could have been further from the truth.

The Pugwash Conferences on Science and World Affairs

Ashok played a major international role in his contributions to the Pugwash movement. The Pugwash Conferences on Science and World Affairs were first started in 1957, in the town of Pugwash, Nova Scotia. The original purpose was to bring together scientists from the East and West to discuss how science might be mobilised to help minimise the risks of weapons of mass destruction. In the mid-1960s, the topics for the conference discussions were widened to include the role of science in removing global poverty. Both Ashok and I were invited to participate in the conference discussions on the latter subject.

Three activities of Pugwash stand out for enhancing our understanding on how science and technology contribute to development. These are the preparation of a code of conduct on technology transfer to developing countries; a review of what is meant by scientific and technological self reliance; and the preparation of guidelines for international scientific collaboration. Each of these topics has strong political overtones and was taken up by Pugwash to try to find scientific and technological solutions to politically contentious issues.

Ashok helped to convene and then participate in the working groups on each of the latter two issues. I will focus on his role in the working group to develop guidelines for international scientific collaboration. This is appropriate, because it was Mrs Gandhi, when she was the prime minister, who asked Pugwash to prepare such guidelines. She did so in the light of several cases of international scientific collaboration involving Indian and foreign scientists, which had been perceived to be detrimental to India.

In some instances, the Indian partners had been used for data collection but the analysis and publication of results had been in the name of the foreign scientists. There were also cases where the 'collaboration' was merely a guise for research that had military implications. For example, there had been studies on migration of birds, which were sponsored by foreign governments and which could have implications for biological warfare. Sometimes there were economic implications of the research, where it appeared that the foreign partner gained most of the benefits.

These and other examples had led the Indian government to restrict approvals that were needed for foreign scientists to carry out research projects in India. India had become almost as closed to foreign scientists at this time, as China had become.

The Indian Pugwash group responded to Mrs Gandhi's request by convening a workshop, just outside Delhi. Ashok was one of the Indian conveners of the event that was attended by approximately thirty scientists from North and South. The objective of the workshop was to consider the problems, and if appropriate, to draft a set of guidelines for the conduct of international scientific research. Although the meeting was held in India and had been requested by the Prime Minister of India, it was intended by Pugwash to be applicable to all international scientific collaboration. The guidelines were intended to apply to scientists, governments and funding agencies and identified the responsibilities and obligations of all the involved parties.

The guidelines were published and distributed to the participants of the 1979 UN Vienna Conference on science and technology for development. I do not know how influential they were on the Conference proceedings but international scientific collaboration today is much more widespread and less problematic than it was prior to the Vienna Conference and the Pugwash guidelines.

Ashok's contributions to the Pugwash debates were considerable. We were not always on the same sides of the debates but we were both open to being influenced by the evidence. The guidelines are testimony of our trust in one another and our ability to work together to solve contentious issues.

Several years later the Pugwash Movement and its Secretary General, Joseph Rotblat, were awarded the Nobel Peace Prize.

There have been many other international events and issues where Ashok made important contributions. He was a strong supporter of the South Centre and helped ensure that India was a member of the Science and Technology Policy Instruments Project. I have only included those events where we worked together.

In short, Ashok not only made many substantial contributions to the development of science and technology policy in India, but he also had a major impact on international science and technology policy. His influence on the survival of SPRU was considerable. On both, the Indian and the international stages, he advocated scientific and technological self-reliance, but his definition of self-reliance is measured.

I have found him persuasive in his arguments, but he has always been willing to modify his position when the evidence warranted it.

Our negotiations were always evidence based and never encroached on our personal friendship.

Ashok has been a science and technology policy researcher, as well as, a policy maker, a policy adviser and a knowledge broker. He has moved between these different activities with ease. Undoubtedly, this ability is one of the reasons why Ashok has had such a successful career. I salute him and his contributions, on his 75th birthday.

13

AMRITA PATEL

Professor Ashok Parthasarathi's Seminal Contribution to the Electronic Transformation of the Nation's Milk Programme

Operation Flood, the programme which led India to become the world's largest milk producer through a network of cooperatives across the country, was based on the model of a successful cooperative that had been set up in Anand, pre-independence. Small farmers brought tiny quantities of milk—two to three litres twice daily—to sell to the village cooperative, where it was tested for its fat content, based upon which, payment was made. Testing of the samples, taken each time a member brought milk for sale, is what enabled members to be paid twice a day, everyday of the year. This payment in cash is what kept the home fires burning and enabled women to pay for their essential daily needs, because other income from land and agriculture was seasonal.

The method for testing fat content in the milk, known as the Gerber test, was cumbersome and tedious. It involved measuring a precise quantity of milk into a glass butyrometer, adding concentrated sulphuric acid to it, followed by a small quantity of amyl alcohol. The top of the butyrometer was then plugged with a metal clad rubber stopper. The contents had to be then thoroughly mixed. A number of butyrometers were then placed in a manually or electrically operated centrifuge. After centrifugation, the fat in the milk needed to be read on a graduated scale which was etched on the butyrometer. This was an expensive process as the chemicals had to be transported over long distances. It also required skill, since it was not uncommon for acid to spill out of the butyrometer. Frequently, the rubber stoppers leaked and the hands of testing staff were exposed to the concentrated acid.

There was clearly a need to replace this method of testing by one which was safer, faster and reliable.

One of the critical interventions in the story of the White Revolution through Operation Flood, was the introduction of the Electronic Milk Tester to test milk at the place where the milk began its journey to the consumer—the village.

At that time, A/N Foss Electric of Denmark was one of the most successful companies in the world that was manufacturing milk-testing equipment. The National Dairy Development Board (NDDB) had in fact ordered 14 lab model milk tester machines from them, for the larger dairies being set up under Operation Flood. These machines were expensive (about US$ 30,000). They operated on the principle of light-scattering, which depended on the fat content of the milk. The effect of other components in milk was eliminated by using a chemical diluent (ethylenediamine tetra-acetic acid—EDTA based), and the fat globules were made uniform by a homogeniser. Clearly, the machines could be used only by trained technicians and were not sturdy enough to be used at the village level.

Around that time, a team from Foss visited NDDB in 1975, to know more about Operation Flood and the demand for milk testing machines. They were taken to the villages to see, at first hand, the conditions under which milk was being tested. The need for a better system for testing millions of samples of milk, morning and evening every day, from lakhs of villages under Operation Flood, was explained to them.

In subsequent meetings with NDDB, the possibility of developing a suitable machine for use in the villages was discussed with the Foss team. Interest in developing a suitable machine led Foss to work with NDDB's product development group in developing a suitable machine. A number of suggestions made by NDDB, such as replacing the automatic pipettes for milk and diluent with manual push button pipettes, and the use of a hand-cranked homogeniser instead of an electronically operated one were incorporated in the design, which made the machine simpler and able to operate with less use of power. This also enabled the use of a rechargeable car battery, as electricity supply in villages was unreliable. Foss and NDDB together worked on the machine with considerable testing at the village level and a workable electronic tester was developed by 1978. Foss was able to reduce its cost to about US$1500.

At this point, NDDB sought the help of Ashok Parthasarathi, who was at that time, Joint Secretary, Department of Electronics, Government of India. Ashok, appreciating the enormous impact the machine could have on the lives of millions of small and marginal farmers in our villages, took up the challenge personally and played an important role in its development thereafter.

Ashok's assistance was sought by NDDB in arranging for mass production and marketing of the Electronic Milk Tester. Ashok decided that the best and fastest path was to set up a joint venture company. With the passion and drive, which are characteristic of Ashok, he persuaded the Ministry of Heavy Industries and Public Enterprises, Government of India, to agree to set up a joint venture in which they would hold 51 per cent shares (through the public sector undertaking, Instrumentation Ltd. Kota, Rajasthan). He then persuaded the Rajasthan State Industrial Development and Investment Corporation Ltd. to be the other partner, to hold 49 per cent shares in the new venture.

The joint venture company, which came to be called Rajasthan Electronics and Instrumentation Ltd. (REIL), came into existence in 1981. Ashok played a key role in policy making and in the smooth running of the company in its early years. In fact the importance he gave to establishing the company was demonstrated when he allowed one of his own trusted engineers at the Department of Electronics to become the first Managing Director of REIL. And so, a high class manufacturing unit was set up under his guidance.

To conclude, Ashok Parthasarathi has indeed made a major contribution in a variety of areas in the field of science and technology in our country. His understanding, commitment and steering of the setting up of REIL, by overcoming the bureaucratic hurdles that are inevitable in setting up of such a venture, will not be forgotten. The Electronic Milk Tester over the years has changed the lives of millions of small farmers in our country and has led India to become the largest milk producer in the world.

14

PRABHAT PATNAIK

My Colleague Ashok

It must have been in 1973 that I had gone with my colleague, Professor Krishna Bharadwaj, from the Centre for Economic Studies and Planning of the Jawaharlal Nehru University (JNU), to listen to a talk that Ashok Parthasarathi was giving on technology policy in China. The talk was excellent. That was the first time I saw him, though I do not recollect being introduced to him on that occasion. I had, of course, heard of him, of his superb intellectual abilities, and of the fact that at a very young age he had become the science and technology (S&T) advisor to the prime minister, a position he held for nine years (in two spells). There was an additional bond: his father G. Parthasarathi (GP) was our vice-chancellor, the founder of JNU, who had been the chairman of the selection committee that had recruited my wife, Utsa and me to JNU. We were in Cambridge before this and reports of this exciting new university coming up in Delhi under GP's leadership had lured us back.

Another close bond with Ashok's family was to develop a few years later when our older son joined the Sardar Patel Vidyalaya (SPV) in Delhi, where Ashok's wife, Vibha, was the principal. Since our son attended SPV from nursery until the end of class XII, this relationship with Vibha was a long and a close one, though, of course, from our side, admiration and affection for Vibha was tinged with an element of awe and even trepidation which all parents feel vis-à-vis the principal of their children's school, being only too aware of their own flaws in parenting.

The early seventies were a period of strong anti-imperialism in the country, in the wake of the Bangladesh war and Nixon considering the use of nuclear weapons against India. JNU itself was a product of this period, which also witnessed a re-emphasis on the virtues of national self-reliance, and a re-assertion of allegiance to what one may call the Nehruvian development strategy. Ashok's own intellectual predilection lay in that direction and he emerged as a major figure in this project, in the realm of science and technology policy.

The sixties, marked by the two wars against China and Pakistan; massive crop failures; inflation of the middle-decade, the Bihar famine (though nowhere on the scale of the pre-independence famines); and devaluation of the rupee under the direction of the World Bank, had been a trying time for the Nehruvian strategy, and a period when American influence had spread considerably. Against this, the last year of the decade and the subsequent years of the early seventies marked a return to the ideological moorings of the fifties. Nationalisation of banks; the Green Revolution; emphasis on control over foreign capital and monopolies; and a re-assertion of the need for self-reliance in the interest of the people, were the chief hallmarks of this return. Alas, the build-up to, and the imposition of Emergency, put an end to these. Ashok was at the centre of things at this time, a staunch adherent of this strategy.

The Jawaharlal Nehru University, which he joined after his stint in the government, was, in a sense, the natural habitat for him. It was nonetheless, a brave move. For persons of his level of seniority to come into the university system is far from easy. The pre-existing faculty looks askance at the newcomer, and for the latter too, the somewhat chaotic and bumbling manner of the functioning of the university comes as a shocking contrast to the ways of the government; the easy-going slowness of its administration taxes one's patience. In Ashok's case too, there were these teething troubles but he adjusted admirably.

He joined the Centre for Studies in Science Policy (CSSP) in the School of Social Sciences, and soon became its chairperson. His tenure as chairperson of the Centre, broadly coincided with my deanship of the School; and I was a witness to his adjustment. Teaching at the CSSP, which had been an innovative venture on part of JNU—believing as it did, in inter-disciplinary and off-the-beaten-track research—received an unprecedented infusion of financial and intellectual resources, when Ashok was at its helm.

During that period, we met almost on a daily basis, perhaps even several times a day, and not only at formal meetings of bodies like the Committee on Advanced Studies and Research, the Chairpersons' Committee, and the Board of the School, which constitute the core of the academic administration of a school in JNU. Ashok's counsel at these meetings, based on his long experience, was always sagacious, and full of sympathy for students (an important requisite in a university). He fitted in extremely well with the progressive liberal ethos of the university. And one remarkable quality of his, which impressed me, was his freedom from malice in any form.

After my stint as the dean ended, I naturally saw less of Ashok. And after I moved to Kerala as the vice-chairman of the State Planning Board, I lost touch with him, until I heard from him again when he was organising a series of panel discussions on the different aspects of GP's life-work, at the India International Centre, on the occasion of GP's centenary. He had asked me to be part of a panel, alongside Romila Thapar, Prakash Karat (a student leader at GP's time with a close relationship with GP), Sukhadeo Thorat, and Neerja Jayal, to discuss JNU during GP's days. The proceedings were taped, transcribed and edited by the authors. They provide significant insights into the history of JNU and also into the social, political and intellectual life of the country at that time.

The proceedings of the six panels and four lectures, together with a lot of supporting material will be published as a book called, "Remembering G.P. in his Birth Centenary Year". This volume is edited by Ashok and Meera Shanker, an Indian Foreign Service Officer, who had worked closely with GP when he dealt with the Sri Lankan imbroglio.

I see this Festschrift as only one more milestone in a long and distinguished career and wish Ashok many more years of active and socially productive life.

15

A.N. PRASAD

In Honour of Ashok Parthasarathi

It is indeed laudable that the friends and associates of my good old friend Ashok Parthasarathi have thought it fit to bring out a Festschrift in his honour, coinciding with his platinum jubilee birthday. I consider it a privilege to make this humble contribution to this Festschrift. I must confess that my personal contacts with Ashok have not been too frequent. But still, I have always cherished knowing him well enough to engage in long conversations on topics of mutual interest, including science policy, whenever we met at conferences, panel meetings, or when we spoke over the telephone. I particularly recall his visits to my residence in Bangalore, where I now live, and the lively conversations we had. He is a really fine gentleman. Some years ago, we both travelled together to California to attend an International Pugwash conference, and I can never forget his wonderful company and the interesting time we had together.

This Festschrift cannot be complete without a detailed account of the persona, the positions and responsibilities Ashok Parthasarathi has handled, his contributions and achievements in his long illustrious career. It will be superfluous on my part to even attempt elaborating it, as I am sure his impressive biodata will figure prominently in this Festschrift. I will therefore restrict myself to highlighting some of the aspects I am familiar with.

Right from the start of his career at a young age, Ashok Parthasarathi has been fortunate to have had increasingly diverse exposure in various fields of national importance. In addition to his core area of expertise, electronics, he was engaged in strategic areas like space, nuclear energy and defence. With his vast experience over the years, in various capacities in the central government and in international agencies, he seems to have now shifted his focus towards issues related to national S&T policy and analysis. Perhaps one of the most visible and significant contributions of Ashok during his decade long tenure in the Department of Electronics (DOE) was that he was closely involved in the planning and promotion of a number of professional, electronic public sector companies, including the Semiconductor Complex Ltd. at Chandigarh. As Secretary, Department of Scientific and Industrial Research (DSIR), he has encouraged small and medium scale industries, in particular, in developing and commercialising innovative indigenous technology, by directly financing the enterprises through the DSIR. He also introduced many measures, such as providing income tax deductions, to stimulate and support enterprises to undertake research and development (R&D). This, I consider, is a very important step for the good of the industry, particularly in the small and medium scale sector. I recall serving in some of the panels set up by DSIR for evaluation of proposals received for innovative indigenous technology awards.

Not many, even among the old timers in the Department of Atomic Energy (DAE), are aware that Ashok served in the DAE from 1967-70 as special assistant to the then chairman of the Atomic Energy Commission (AEC), Dr Vikram Sarabhai. In that position he did many things. First, he did a lot of work for Dr Sarabhai relating to the entire nuclear energy programme itself. Second, he set up and headed the Programme Analysis Unit, that provided policy planning and management services to the Atomic Energy Commission. And third, he worked as a senior staff member of the Secretariat of the Electronics Committee of the Government of India, which Dr Sarabhai chaired. As for the nuclear programme, his interest was not confined to the energy aspects of the programme alone. During my meetings with him and our long conversations, I could sense his strong commitment to keeping abreast of developments taking place in the nuclear sector vis-à-vis their impact on national security as well.

Since I had fully dedicated my entire professional career to the nuclear programme, which is close to my heart and soul, I am tempted to take the liberty of expressing some of my views and feelings on recent developments that I am concerned about. As those developments were of concern to Ashok also, my doing so, may not be out of place.

Contrary to all the negative publicity and apprehensions expressed in some quarters, there is absolutely no doubt in my mind that, in the long run, the nuclear energy sector is bound to make a significant contribution to global energy needs. It is a clean source of energy devoid of the adverse effects of green house gases and global warming. Of course, the health, safety and security concerns have to be meticulously and comprehensively addressed. It is unfortunate that the natural evolution of nuclear energy in the world has been badly hampered due to fear of nuclear proliferation. This has resulted in a steep reduction of its full potential for societal benefits. Dr Homi Bhabha, a great visionary and founding father of our nuclear programme, always used to advocate international cooperation in the peaceful use of nuclear energy for the good of humanity. However, being a realist, to effectively safeguard our national interests against the vagaries of international power politics, he was committed to building robust and self-reliant indigenous capability and capacity, encompassing all aspects of the nuclear fuel cycle to ensure two goals: long-term energy security and national security. Based on this premise, during the last nearly six decades, we have systematically developed and put to use technologies encompassing the entire nuclear fuel cycle. For quite some time now, we have been able to design, engineer build, operate and maintain nuclear power reactors, completely indigenously. Much of this capability was achieved owing to our compulsions of having to undertake the whole programme on our own—defying embargoes and innovating around technology denials from supplier countries—in the aftermath of our nuclear tests, first in 1974 and then in 1998. As a result, and because successive governments have refused to sign the discriminatory Nuclear Non-Proliferation Treaty, we are now seen as an advanced country in the nuclear field.

Recognising that reality, the United States government perhaps thought it fit to engage with India and enter into a civil nuclear cooperation agreement. The proposal first mooted in 2005, by both the countries jointly, seemed to be a historic move towards ending some thirty years of nuclear apartheid and India coming out of severe externally imposed isolation. It was anticipated that this would also open up avenues for India's collaboration with the outside world to accelerate our nuclear power programme, which has been growing at a very modest pace, mainly due to successive chairmen. AEC had not given top priority to exploration and exploitation of new deposits of uranium, which could have helped India go around western trade embargoes on uranium and practically anything nuclear. The agreement with the US government was also expected to provide opportunities for us to share our expertise and play a truly global role in the provision of safe nuclear energy.

When, after three years of intense negotiations, the so-called Indo-US 'Deal' was finally signed in 2008, it was stated to be a "Full Civil Nuclear Cooperation Agreement". However, detailed analysis of the Agreement revealed a significant omission in its scope. When an Agreement is called "full", it should obviously cover the entire nuclear fuel cycle. However, there was a big lacuna. Two crucial links of the nuclear fuel cycle, that is, Uranium enrichment and spent fuel reprocessing, together called ENR technologies, were clearly excluded from the scope, thus rendering the Agreement in effect 'partial' though touted incorrectly as 'Full'. ENR technologies are important for achieving and sustaining our long-term nuclear power programme on a self-reliant basis, and their exclusion could slowdown the pace of the stages of that programme.

Also, if one goes through the fine print, there are anomalies in the terms of the Agreement showing lack of a level playing field among the two 'partners', that is, India and the USA, and far too many compromises on our side, under pressure from the USA. It appears the Agreement is heavily biased towards those parts of the nuclear fuel cycle which are of commercial interest to the supplier countries. I have elaborated on these in many of my articles, media interactions, debates, panel discussions and talks in the past.

The sad part is, when the Agreement was being worked out, totally impractical and unrealisable targets for generation of power, for the years 2020, 2030, 2040 and so on, were being freely rolled out to make the proposal (which was facing flak from many quarters here) acceptable to the public. It is a pity that, even after seven years of signing the Agreement in 2008, not a single MW of power has been generated under this arrangement! We are still groping with issues related to nuclear liability and land acquisition and it is not clear how they will plan out. The fine print of the provisions of the liability issues, as worked out by the new government recently, are still kept under wraps and not made public, leading to speculations. However, informal discussions with the negotiators, in both the DAE and the Ministry of External Affairs, have indicated that the problems with our Civil Nuclear Liability Law remain on the side of the USA and France. It is high time some credibility is restored in the implementation of the Agreement. One other important fallout from this Agreement is that, for all practical purposes, there can be no more nuclear tests. To what extent this will impact national security *vis-à-vis* our neighbouring countries, is a big open question.

In dealing with the topic of the Agreement, I have tried to vent my feelings, even if they sound somewhat critical and even pessimistic. There is no point in shying away from ground reality. Having devoted my entire professional career to the nuclear establishment and its development, at my present age of 80 years, my only wish is that, the DAE, which I have served with dedication, should continue to maintain its full glory in the years to come, contributing to the good of the nation. It hurts when hard earned scientific and technological expertise and self-respect are compromised. There is so much to introspect about and still a lot more to be achieved.

Let me end by wishing Ashok, through this Festschrift, all the very best in the years to come and pray that he continues to contribute to the betterment of the country, as he has been doing all these years.

16

FRANCISCO SAGASTI

Remarks in Honour of Ashok Parthasarathi

I had the privilege of meeting Ashok Parthasarathi in the early 1970s, when I became the field coordinator of the Science and Technology Policy Instruments (STPI) project that gathered ten developing countries to study the design and implementation of policies to promote science and technology for development (see the tribute by Geoffrey Oldham in this Volume). At that time, Ashok, though in his early 30s, was the science and technology Adviser to Prime Minister Indira Gandhi, and his good offices secured the participation of India in the STPI project. I remained in frequent contact with Ashok during the 1970s and early 1980s, and had the opportunity of benefitting from his wisdom and experience, particularly during the negotiations of the United Nations Conference on Science and Technology for Development held in Vienna in 1979.

One of the most memorable occasions I had to work with Ashok arose out of his involvement with the Pugwash Conferences on Science and World Affairs, which were founded by Bertrand Russell and Albert Einstein. Ashok was a member of the Pugwash International Council and one of the organisers of a symposium on self-reliance and development, held in Dar-es-Salaam, Tanzania, in June 1975.

The 1970s were a period of major changes at the international level, prompted by the decolonisation processes: the rise of the Japanese economy; the beginning of economic reforms in China; the sudden increase in oil prices in 1973 and again in 1979; and the creation of the Organisation of Petroleum Exporting Countries (OPEC), among other forces. These led to a questioning of prevailing ideas of development, a realignment of international power relations, and to the launch of the North-South Dialogue in Paris, which was supposed to debate and agree on proposals for a New International Economic Order (NIEO).

At that time, there were several attempts to articulate alternative conceptions of development. Ideas such as "redistribution with growth" (Chennery), "circular cumulative causation" (Myrdal), "basic needs" (Singer, Streeten, Jolly), "unequal exchange" (Samir Amin), "small is beautiful" (Schumacher), and "eco development" (Sachs, Strong), among many others, aimed at providing guidance for designing and implementing development strategies. In this context, 'self-reliance' emerged as a powerful concept that explicitly sought to avoid the pitfalls of imitative development. While some viewed self-reliance as a variant of autarchy, more sensible and moderate interpretations emerged to take centre stage in development debates. Ashok was a key figure in formulating the interpretation of self-reliance that prevailed in the end, and which still maintains its relevance.

Ashok co-edited the volume *Pugwash* on *Self-Reliance* (Chagula *et al.*, 1977) gathering the papers presented at the Dar-es-Salaam symposium. His summary report, *The Role of Self-Reliance in Alternative Strategies for Development*, (Parathasarathi, 1977) stated clearly: "A development strategy imitative of highly industrialised countries appears to be neither possible, necessary or indeed desirable for the majority of developing countries." Resource constraints, consumerism, environmental degradation and alienation were among the main reasons for rejecting such a path, explicitly advocated by development economists like W.W. Rostow in his book, *The Stages of Economic Growth*.

All of us who participated in the Pugwash Dar-es-Salaam symposium agreed on the definition provided by Ashok Parthasarathi (1977: 14): "Self-reliance is to be understood at the national level of each developing country as the will to build up and use a capacity for autonomous decision-making and implementation, on all aspects of the development process including science and technology." Autonomous decision-making in matters of science and technology was also a key issue identified by several Latin American colleagues of Ashok, including Jorge Sabato and Amilcar Herrera, whom he knew well. I had the opportunity to interact and learn from all of them, particularly within the framework of the STPI project, and my exchanges with Ashok were most helpful to shape my own ideas on how to mobilise science, technology and innovation for development.

In a prescient way, Ashok's Dar-es-Salaam report highlighted the crucial role of indigenous capacities to generate and utilise scientific and technological knowledge, develop domestic technological know-how, identify and import foreign technologies on the best possible terms after effective negotiation with the technology supplier and blend indigenous with imported technologies in the most appropriate ways, taking into account local contexts. These ideas have become even more important as the challenges of the knowledge society have become a pressing reality for all countries in the first decades of the 21st century.

Throughout his long and prolific career, Ashok Parthasarathi has shown an unusual ability to combine outstanding intellectual qualities and leadership, with active policy-making and public service. He has been an inspiration for many of us who have tried to emulate his achievements. It is a great honour to be one of his numerous friends and admirers, and to join a celebration of his life and accomplishments.

References

Chagula, W.W., B.T. Feld and A. Parthasarathi (eds.) (1977). *Pugwash on Self-Reliance.* A Pugwash monograph based on the Symposium, held in Dar-es-Salaam, Tanzania, 2-6 June 1975. New Delhi: Ankur Publishing House.

Parthasarathi, A. (1977). "The Role of Self-Reliance in Alternative Strategies for Development", in W.W. Chagula, B.T. Feld and A. Parthasarathi (eds.), *Pugwash on Self-Reliance.* A Pugwash monograph based on the Symposium, held in Dar-es-Salaam, Tanzania, 2-6 June 1975. New Delhi: Ankur Publishing House. p.11.

17

L.K. SHARMA

An Unusual Technocrat

Ashok Parthasarathi, an unusual scientist-civil servant, made significant contributions in the area of science and technology policies and programmes for around 40 years, from the late 1960s to the first decade of the 2000s. In the decision-making bodies, task forces and committees or in advice to the prime minister, Ashok expressed his views very effectively. He could do so because of his deep understanding and experience of the issues related to the promotion of science and technology and their application to the development and security of the nation. He also considered science and technology in the wider context of our foreign policy and strategic interests as well as in their social and cultural aspects. He studied the linkages between science and technology in great depth. He also never overlooked the policies of the big powers to curtail our competence.

Ashok came into the government after studying and researching in physics, astronomy and electronics on the one hand and S&T policy-making and policy research on the other. He spent a lot of time every day to study the hot topics calling for decisions and constantly updated himself in his chosen areas. His rigorous research and mastery over details made his inputs for decision-making invaluable. However, as was to be expected, this unique capability of his did not endear him to some of the generalist bureaucrats, who constantly queered the pitch for Ashok in the inevitable bureaucratic battles. But he was able to deal with such obstructionist efforts effectively and won most of the battles.

The issues relating to make-or-import technology, technology selection, technology trends and allocation of resources to S&T, in particular, are always complex. These are muddied further by the vested interests trying to distort government's policies and decisions. All his working life, Ashok fought to defend national interests, continuously contributing to technological self-reliance and frustrating the designs of those seeking to undermine it.

In the pursuit of this objective, Ashok did not play safe, nor avoid controversies. Because of his ideological commitments, questioned by the vested interests, Ashok got incorrectly categorised in New Delhi's '*babu*dom' as an obstructionist. By following the principle, know your enemy, he became an expert in thwarting vested interests. Ashok had the knack of collecting sensitive information in the areas of defense and development and using it to give decisive strength to his arguments.

When Ashok took a stand on a particular issue, after a thorough study of it, he stood firm and most often won the day. He did not modify his opinion even when he saw the wind blowing the other way. This exposed him to the charge of arrogance but it did not affect his commitment to his drive for self-reliance. His mastery over details won little admiration in the official meetings and provoked envy, particularly because such mastery(ies) enable him to win the day.

At times, a driven bureaucrat has to pay for his uncompromising nature. This is not to say that Ashok a technocrat par excellence was always right. Technology development, the country's core competence and industrial culture are factors that present a set of uncertainties and the best futurist may subsequently find that his calculations were wrong. When that happened, Ashok had the humility to accept his mistakes.

Ashok was able to overcome problems and cross hurdles, despite numerous difficulties that affect the functioning of decision-makers when they are up against vested interests and lobbies that operated at the best of times.

Fortunately, in the era in which he operated in New Delhi, Ashok enjoyed strong and sustained political support from late Prime Minister Indira Gandhi, whom he served as science and technology advisor for nine years (over two spells). His strong commitment to technological self-reliance cohered with the ideological stance of Indiraji. He had the good fortune of being in the Prime Minister's Office when Indira Gandhi took numerous initiatives across the whole spectrum of science and technology, most of which were at the instance of Ashok. Time and again, Nehru's daughter demonstrated personal commitment of a high order, to shaping the nation as a major scientific and technological power.

At the international level, Ashok's interests covered North-South relations, South-South Cooperation, international trade regimes, intellectual property rights, research and development (R&D), nuclear non-proliferation, nuclear energy and renewable energy. He used his large personal library of books, journals and newspaper clippings on these subjects. He closely watched over the public sector companies and nurtured and protected many of them through radical decisions and personal interventions. He had the advantage of having worked closely with scientists, such as Vikram Sarabhai, former Chairman, Atomic Energy Commission; M.G.K Menon former Chairman, Electronics Commission; and Satish Dhawan, former Chairman, Space Commission, as well as, top administrators, such as. P.N. Haksar, Principal Secretary to Indira Gandhi; and T. Swaminathan, Cabinet Secretary.

Any words written in appreciation of Ashok's contributions, must be written in the context of the India that was Ashok's, which blossomed as flowers of freedom and became the causalities of the 'new India'. Today, Ashok is a person within whom different stages of history clash and the two India's—past and present—collide.

Those cherishing the old idea of India, find themselves marginalised, if not ignored. They cannot fight. They somehow hold on to their unfashionable views on secularism, technological self-reliance, foreign and strategic policies, and an appropriate model of economic social and cultural development. This large community, including the movers and shakers of yesterday, has been relegated to a large retirement home.

If Ashok had poetic sensibility, he would recite:
Waqt teri yeh ada main aaj tak samjha nahin;
Meri duniya kyon badal di mujh ko kyun badla nahin.
(Time, I have not fathomed your ways.
Why did you change my world but not me!)

18

V. SIDDHARTHA

Present at the Creation

I have had to think a long time about what to say here and even longer about how to say it. Ashok Parthasarathi (AP) and I, a little younger, are near-Midnight's Children—a 'doomed generation', as Nehru labelled us—meaning a generation whose sweat and tears would result in rewards only to the generations which came after.

The post-colonial legacies of historical, international and national contexts, as also the collateral domestic political economy in which an establishment of science and technology formed and contributed into the 'second wave' of development in science and technology in our country, are outlined in the contributions by Chakravarty, Sagasti, Oldham, and Roddam Narasimha, to this Festschrift. The height, speed, force and structure of the wavelets of that second wave have been extensively described in AP's seminal book: *Technology at the Core: Science and Technology with Indira Gandhi* (hereafter, AP's book), which is the story of the ways in which AP applied his conceptual and bureaucratic skills to further the cause of technological self-reliance.

This essay is about the beginnings of a 'third wave' in the development of science and technology in India, in which I was privileged to participate. It was a wave which began to form when Indira Gandhi came back as prime minister in 1980 and called for several of the case files that AP had handled. She also recalled, I am sure, her own rhetorical question on Pokhran-I to him, recorded as the last sentence in AP's book: "How do we change the mind-sets of those in the West who are supposed to be our friends?"

This essay is also written for those who know the Bangladesh war only as it is described in school history books; who were toddlers at the time of our first nuclear test in 1974; and who are now occupying the driver's seats in the science and technology agencies in our country, in our industrial enterprises, and in our leading institutions of technical education.

But first a lament: AP is blessed with a prodigious memory for event, time and occasion—and is disciplined enough to have maintained a diary. But the warp and weft of the second and third waves in the tapestry of the development of our science and technology, are lost with those heads of agencies and institutions who were privileged to be 'present at the creation', but who have not left us their record, or recall even, of the events and circumstances, national and international, which informed the decisions in their time: ISRO's Dhawan; CSIR's Sidhu; ICMR's Ramalingaswami; Vikram Sarabhai Space Centre's Brahm Prakash; Raman Research Institute's Radhakrishnan; Hindustan Aeronautics Ltd's Raj Mahindra; National Research Development Corporation's Sharma, and many others.

There is an urgent need to add a 'National S&T Archive' to the National Science Library in New Delhi. A repository that houses a complete digital copy of all publications from S&T agencies; from the research and development (R&D) divisions of Public Sector Enterprises; from co-operative research organisations in the private sector, as also the administrative files of S&T agencies, which are statutorily archived in the National Archives in New Delhi. Lodged and preserved here, should also be an archive of oral history—taped interviews and conversations with the (now ageing) dramatis personae who were participants in the momentous events in our S&T history, particularly post Bangladesh war (1971).

NCST Days

AP's book has an extensive insider's account of the mandate and work of the seminal National Committee on Science and Technology (NCST). But there is an unmentioned episode that had—in my judgement—a far reaching consequence: the derailment of the then-incipient technology policy relating to our industrial development.

After my doctoral work at Imperial College in London, in 1971, I had completed a year as a research fellow at the Science Policy Research Unit of the University of Sussex (cf. Geoffrey Oldham in this Festschrift), when Mr G. Parthasarathi, the founding vice-chancellor of the Jawaharlal Nehru University (JNU) invited me to take a faculty position at the new Centre for the Study of Science Policy at JNU. I was happy to agree, of course, but discovered after my return to India, that other plans had been hatched for my future by him (GP), his son AP, and C. Subramanium, Minister for Planning for S&T. "I think you should join the government" said GP, and I was inducted as the principal scientific officer in the secretariat of the NCST, at that time, an exalted designation and position in the government for someone only in his late-twenties.

I got to work with AP on drafting the Approach to the Science and Technology Plan. The pre-final draft that went to the NCST for its consideration and approval had a end-section titled 'References' which listed the UN and other documents with which AP himself was closely associated, and which had been consulted in the drafting process (cf. Sagasti in this Festschrift). AP argued that the Approach was not an academic paper, but a policy document, and that no policy document of the government had references. I pleaded otherwise. AP's view prevailed.

Through a process described in extenso in AP's book, the Approach was approved by Indira Gandhi and her Cabinet, and released for wide circulation in mid-January 1973. AP records (p.51): "When I told her about the critics, her prompt reaction was, 'I am not surprised'".

As AP's book records on page 52, in her winding-up address to the National Development Council on 19 January 1973, Indira Gandhi noted: "A distinctive feature of the Fifth Plan (1969-1974) is that we are going in for a science and technology plan as an integral part of closely related with and indeed derived from, the Fifth Development Plan for the very first time in our history."

Once it thus became clear, that the NCST's Approach document and its policy prescriptions were serious business, and it was not just another government paper of no great consequence, a media-driven diatribe was instigated against the Approach on alleged impropriety in un-acknowledged use of source documents, charges of 'plagiarism' were flung about, and the whole exercise sought to be discredited.

The real target of the diatribe was AP (I was caught in the cross-fire, but that is another story).

While Minister C. Subramanium handled a messy political fall-out from the above controversy, adroitly, it left the S&T community more-or-less confined to its 'supply side' territories of scientific agencies, institutions and laboratories, as also their funding; while the economics fraternity disconnected from it—nay, disdainful of it—animated the 'demand side', a mismatch, if not a stand-off, from which the planning process of the Planning Commission has been bedevilled ever since.

Incidentally, the largest receptacle for R&D monies of the government—Defence R&D—has remained exempt from that stand-off, despite it being of great consequence for the national S&T effort, because the outlays on it are an integral part of outlays on 'defence', and all of these are 'non-Plan' items, and hence, outside the purview of the Planning Commission. What fortuitous absurdity!

Had a Technology Policy Resolution been adopted by Parliament at that time, as proposed in the NCST Approach document, and had the collateral institutional been built and steps been taken to follow-through such a Resolution, the impact would likely have been as significant for India's self-reliant industrial development, as the impact of the famed MITI and its policies in the fifties and sixties were, for Japan's industrial resurgence after its devastation in World War II.

Atomic Energy and Space

Unlike AP, my post-NCST connect with the Department of Atomic Energy had been indirect, limited to occasionally handling papers of the Atomic Energy Commission (AEC) when I was in ISRO and Satish Dhawan, its chairman, was a member of the AEC, and through invited talks given to its scientific cadres, and professional visits to its facilities. AP's book relates the story of the decision-making on our first nuclear test in 1974. The yield of that test has been variously estimated and remains a matter of speculation and controversy, albeit of no consequence for our current nuclear weapons programme, or its credibility.

Post-shot radio-chemical analysis is the confirmatory method of estimating the yield of an underground nuclear test. In the foyer of the auditorium of the India International Centre, a respected science correspondent of a news magazine and I had the opportunity to ask the late P.K. Iyengar (former chairman of the AEC who was deeply involved with the design of our first bomb) about the results from radio-chemical analysis of the bore-hole samples from that test.

"That never worked," said Iyengar, in an accent characteristic of those who come from the border regions of Tamil Nadu and Kerala.

AP's book describes the mess that Sarabhai left ISRO in. The agency and its programme were rescued by the duo of Dhawan and Brahm Prakash. The story of how that was accomplished would fill several books, but one incident relating to Sarabhai bears mention here.

At the National Institute of Advanced Studies (NIAS) in Bangalore, the late Raja Ramanna, who was the founding director of NIAS and also ex-chairman AEC, and I were listening to a eulogy on Sarabhai by a self-confessed acolyte. After listening for an excruciating time, I whispered to Ramanna: "I never worked with Sarabhai, but I get the impression he was an over-rated man." "Very much," said Ramanna, "very much," clearly relieved at being unburdened from the discomfort he was feeling at the eulogy.

Renewable Energy

Almost as soon as he was appointed secretary of the Ministry of Non-Conventional Energy Sources (now Ministry of New and Renewable Energy), AP press-ganged me to chair a group to detail Indian Standards for solar-powered water-pump installations. And then to chair the Research Advisory Council of the Centre for Wind Energy Technology (C-WET) in Chennai—a position I was requested to continue in by his successor in the Ministry, even after I had formally retired from the government.

At the National Commission for Scheduled Castes and Tribes

It would be politically incorrect to say that AP was for a time 'relegated' to the position of secretary of the National Commission for Scheduled Castes and Scheduled Tribes. But in that role as well, he applied his mind and administrative skills to formulate and execute plans that applied S&T in ways that served to improve the livelihood and welfare of these socially disadvantaged groups. I know, because he called me to his office to ask if I had any ideas about how to locate and use technologies, which could generate employment for these socially disadvantaged groups.

Completely *au fait* with the workings of government, AP has continued well into retirement his avocation—the closest description for which is 'technological intervener'. In that role, I am not sure he is always welcome in the sandstone corridors of Lutyens' New Delhi, but why will that obstruct my friend-in-cause, Ashok Parthasarathi?

19

RAHUL SINGH

Ashok in Our Years at Cambridge University, UK, as I Remember Him

Both Ashok and I joined the Cambridge University at the same time, in 1960. I had just turned 20 then and Ashok was a few months older. He already had a Bachelor's degree in Physics from the University of Madras, while I did not. Hence, he did his Cambridge degree in two years (which was common for graduates from here), while I took three years. I am not sure how we first met at Cambridge but we seemed to have hit it off, right from the start. During my years at Cambridge, I regarded him as my closest friend and we spent a great deal of time together, though we were pursuing different disciplines: Ashok, science and I, history.

We also came from different backgrounds. I was a Sikh, with a turban, beard, and long hair, Ashok was a balding Tamil brahmin/parsi mixture from Madras (now Chennai). Ashok was an extremely earnest person and took his studies very seriously indeed, and I did not—at least not as seriously as him (though much to his, and my, surprise, we somehow ended up with the same degree—a creditable 2-1, which means first division, second class, just a notch short of the best one could get, a first class). Instead of academics, in my first year at least, I was much more seriously involved with tennis, a passion of mine. I was also a little homesick then and tennis, along with the friends I made with other students who played the sport, made me feel happier on the campus. To my delight, I managed to get into the University team right away in my first year, partly because the team was not particularly strong that year (we lost to Oxford University in the traditional Oxford-Cambridge match). I recall Ashok admonishing me for spending too much time playing tennis and not enough on reading and lectures. But it had little effect. My philosophy was "enjoy yourself while you can", unlike Ashok's single-minded focus on academics. There was also a puritanical streak in Ashok, certainly not in me.

However, I should add that Ashok was a pretty good sportsman himself, in cricket and table tennis. If I recall correctly, a shoulder problem prevented him from excelling further in cricket. But he played table tennis at Cambridge, which he did very well at, getting a 'Blue' (highest rating), representing the University and defeating the Oxford team three years running. This was really not surprising as he had played table tennis for India as far back as 1956. I also played the game fairly well, though not at Ashok's much higher level. I managed to get into the B team of the University and we both used to travel together to play against other teams and universities.

An 'oddity' stands out in my memory from those early Cambridge days. Ashok's official name at the college was 'P. Ashok'. In other words, he gave 'Parthasarathi' as his first name and 'Ashok' as his surname! I remember asking him why. His reasoning was perfectly logical. In Britain, unlike America, except by your close friends, you were always called by your surname. The college porter, who came to know every student in the college, for instance, would address me as 'Mr Singh', as would my professor. And 'Mr Parthasarathi', would clearly have been too much of a mouthful for the Brits, whereas saying 'Mr Ashok' would have been easier. So, in his Cambridge college and university records, Ashok's name was registered as 'P. Ashok' till he completed his first degree! However, he then did a 'deed Poll' changing his name back to Ashok Parthasarathi and it was in that name that he got his second degree—an MPhil research degree in Radio Astronomy, from the internationally famed Carendish Laboratory.

Ashok met his wife, Vibha Nayak, a Gujarati from east Africa, at Cambridge. She was studying geography at one of Cambridge's girl's colleges, Newnham. She was an accomplished classical Bharathanatyam dancer and was performing at one of the venues at Cambridge which Ashok and I frequented. Ashok was riveted by Vibha. Smitten, he could not keep his eyes off her throughout the performance. It was instant love, all of 54 years ago! Ashok's mother came down to Cambridge and hosted an engagement party for them at the Garden Hotel, at that time probably the fanciest establishment in the city, by the river Cam.

Ashok went on to MIT—to the 'other' Cambridge in USA but was lured back home by famed cosmic ray physicist, and chairman of the Atomic Energy Commission, Vikram Sarabhai. Sarabhai was perturbed that many of our best scientific brains were being lost to the West. Among them, of course, was Ashok. Sarabhai got him back at a senior position in the Department of Atomic Energy, which, I later learnt, caused considerable annoyance among the IAS (Indian Administrative Services) set-up there, since he was given a more senior position than civil servants of his age. Ashok eventually became scientific adviser to Prime Minister Indira Gandhi.

But Rajiv Gandhi, when he took over after his mother's assassination in October 1984, developed a strange antipathy towards Ashok, which I have been unable to fathom to this day, but which will probably be explained by others in the know, in this Volume. I can only hazard a guess. Rajiv and Ashok were at Cambridge around the same time, Ashok, who could not suffer fools, knew some of Rajiv's close friends and did not think much of them. And Ashok's blunt nature was such that if he thought poorly of somebody, he did not disguise his feelings. Unfortunately for Ashok, some of these buddies of Rajiv were around him when he became the prime minister. Hence, they were in positions of power. I suspect they had their daggers out for Ashok.

Be that as it may, I don't think Ashok was cut out to be a civil servant. He belonged to the academic-cum-scientific world. He came back home, mainly because, in an old-fashioned way, he felt it was his patriotic duty to serve his country as best as he could. Vibha, too, instead of taking the easy way out, chose to join a struggling Hindi-medium school in Delhi, the Sardar Patel Vidyalaya, and devoted her life to teaching and to furthering the empowerment of women. She soon became the principal of the school and made it one of the finest establishments in the country. My niece and many of my friends' children went to school there.

Ashok and Vibha made their base in Delhi, while I stayed on in Bombay, after I got my first job in that metropolis as an assistant editor with the *Times of India*. So, we did not see as much of each other, as we would have, if we lived in the same city. Nevertheless, when Ashok came to Bombay, or when I went to Delhi, we would invariably contact each other and pick up the threads. That, I believe is the essence of true and lasting friendships: no matter how long the separation between two people, once the connection is re-established, the years of separation immediately melt away.

As I said earlier, Ashok and I were, and still are, very different personalities, in many ways. He took life, and himself, much more seriously than I did. I think he disapproved of my flippancy and light-hearted attitude towards most things, including work. He had strong likes and dislikes, whereas I took everything, including people of all types, in my stride. He was less tolerant towards those who he felt were undeserving of his time and attention, than I was.

Yet, I believe, we both had, and have, certain common traits that in a strange and profound way brought us together in the first place and have bound us together ever since. One was honesty and a sense of decency. I could not see Ashok doing anything dishonourable and I would hope to think that he thought of me in the same way. Perhaps one could call them Gandhian values, since both Ashok's family and mine had been profoundly influenced by the Mahatma. Another common characteristic we had was disdain for pure material gain and goods. Perhaps it helped that both Ashok and I came from fairly well off, upper middle-class families, but the pursuit of wealth for its own sake, was never part of our genes. It was more important to serve one's country. All this may sound a bit pompous and self-righteous, but I cannot put it in a better way. And these values came essentially from our upbringing, from our parents, furthered by our education at Cambridge. Ashok's father and my father, like Ashok and me, were also very different. One, a distinguished journalist-cum-diplomat from perhaps Tamil Nadu's best-known family, and the other a struggling writer whose father had been virtually illiterate (but a successful contractor who was able to provide his children with the best education). Yet, both fathers, Ashok's and mine, despite the vast gulf in their backgrounds, had much the same sense of honour, of selfless duty to the nation, which I would like to think, they passed on to their children.

20

GEORGE SUDARSHAN

My Interactions with Ashok Parthasarathi Over Forty Years

I met Ashok for the first time in New Delhi, in the mid-1970s, when he was science and technology advisor to the then Prime Minister Indira Gandhi, working out of her secretariat. I was very pleased to meet him, since unlike the other scientists I had met and talked to so far, he was more optimistic and enthusiastic about the possibility of accomplishing things in India. For some time, I had been involved in international cooperation projects in the scientific field and had met many Indian scientists, including Raja Ramanna, M.G.K. Menon and M.S. Swaminathan. Of course, I had known Homi Bhabha but at a much earlier stage in my life, as a graduate student at the Tata Institute of Fundamental Research (TIFR). However, I was fully aware of Bhabha's vision for science in the country.

On the several occasions when I met Ashok during my visits to India, I was delighted to have free and frank exchange of ideas on a wide range of topics such as: building a scientific community of high calibre in the country; the nature of the institutions which would help and support scientific endeavours, science policy; giving scientific advice to the government to help create such institutions; methods of improving the research and development practice in the country; and ways to utilise the Indian scientific talent available abroad; etc. I also had the opportunity to meet Prime Minister Indira Gandhi and discuss with her the possibility of scientific work in India compatible with international standards.

During my early days of research at TIFR, I was able to meet outstanding scientists, Nobel Laureates, such as Sin Itiro Tomanaga, Maria Goepert Mayer, Paul Adrian Maurice Dirac and Robert Marshak. Later, when I was abroad, I also met many other well-known scientists, including H. Yukawa and Werner Heisenberg. I had ample opportunity to observe not only their creativity in action but also their methods of fostering and nurturing creativity in their students. This stood me in good stead in my own suggestions to the Government of India for their policy making on science. Exchanging ideas with Ashok and finding that we had similar views, was indeed very encouraging.

While Ashok Parthasarathi has been very active and fully involved in leading, directing and nurturing institutions of applications of science to technology, his commitment to the promotion of basic science is evidenced in his efforts at mobilising funds for that unique institution, the Centre for Philosophy and Foundations of Science (CPFS), from personal and government sources. India's hoary past has always been concerned with philosophical reflections and it has been a world leader in deep philosophical thought. The trend in modern times has been to place much greater importance to developments in science and technology, worldwide.

However, the developments in basic science have reached a stage where researchers recognise that there are very strong parallels between philosophical and scientific conceptions. Under these circumstances India is very much at an advantage in pursuing the philosophical foundations of science which is one of the goals of CPFS. As chairman of the governing body of CPFS, of which Ashok was a member, I appreciated Ashok's help and commitment to the maintenance and development of CPFS very much.

What I found most positive about Ashok was the way he talked about what was achievable. It was always a pleasure to talk to him. His training abroad had broadened his vision and his optimistic nature made him look more enthusiastically at the possibilities, and not be deterred by the difficulties. In my conversations with him, this optimism was very striking and encouraging.

Over the last 40 years, or so, Ashok Parthasarathi has made signal contributions to the growth and development in large areas of science and technology of India.

I am very pleased to contribute to this volume, honouring his achievements, and I wish him many more years of active and fruitful life.

21

P. SUDARSHAN[1]

Ashok Parthasarathi—Outstanding Nationalist, Global Indian, Modern Techno-Managerial Professional cum Development Policy and Governance Expert

Forty-five years have flown since AP interviewed me at the campus of the Indian Institute of Ahmedabad (IIM-A), along with the equally iconic T.N. Seshan, for my very first job. It was for a role in the nuclear programme to promote the applications of isotopes and radiation. I would be a member of the Programme Analysis Group (PAG) located at Mumbai, the Headquarters of the Department of Atomic Energy. I learnt later that AP was himself instrumental in creating and building up this unique multidisciplinary group comprising entirely of 'dual-background' professionals and was the Founder Director. The members of PAG typically had a Science/Technology plus Economics/Management background, a rare mix in the early 1970s. Dr Vikram Sarabhai, then chief of the nuclear programme, since the tragic, untimely loss of Homi Bhabha, interviewed me in IIM-A (yet another creation of Vikrambhai himself) and gave the nod to my appointment. My start in the nuclear programme and my stint with PAG, were rare, holistic, and formative influences for my entire career ever since!

---

1. It is a rare honour and privilege to a life-long professional like me to have been asked to write about a great, noble, patriot, one of my own gurus along with Vikram Sarabhai, Satish Dhawan, Brahm Prakash, Yash Pal, Abdul Kalam, T.N. Seshan, V. Siddhartha ('Sid'), and 'Santy' Santhanam as well! They, Ashok and my great parents, shaped my life and entire career, beyond reckoning, enriching it for decades till this day, always with fond memories of their association and an undeserved kindness, while mentoring me! Ashok Parthasarathi (AP) has a very special place in my career and life.

Ashok soon got a higher calling in the Prime Minister's office as the PM's Advisor for Science and Technology (S&T) matters working in the Prime Minister's Secretariat itself. This position enabled Ashok and his staff to oversee practically the whole of the Central government and the State government as well. Ashok is the only person to have held such a position in our history. Yet, the close professional and personal relationship has lasted to this day. His unique ability to connect with fellow-professionals of all backgrounds, streams and vintages, and work with them on diverse aspects of national development, sets him apart. Persistence and tenacity are hallmarks of his personality. The number of professionals whose careers he has influenced, shaped and developed over more than five decades, is legion!

He is well-known for the courage of his convictions, even while fighting seemingly lost battles. The National Silicon Facility, proposed to be set up by the Indian Petro Chemicals Corporation (IPCL) at Baroda which no less than the Prime Minister herself had called "The Steel of the 21st Century", and the setting up of which she had assigned to AP is a case in point. (The opportunity the country lost in hi-tech production facility garnering a vital, critical, strategic material is rued in saner quarters and corners to this day! It may never come back, not even at a hundred-fold cost today!) Ashok's clarity of vision towards the long-term for the country and his instinct for pouncing on and pursuing strategic opportunities without loss of time, is amazing and has been proven time and again! His erstwhile critics were legion, both petty and professional types. Today, probably they would themselves acknowledge his accuracy—and wisdom, his grasp of the essence, on matters of national importance in science, technology and strategic planning and management! While he is a master in garnering wide, holistic support, he often had to plough a lonely furrow to pursue his convictions.

Throughout his career in Delhi, Ashok, with a strong background in Science and not belonging to any of the central services such as IAS or IFS, had to constantly steer through the mazes and mire of a vast politico-bureaucracy, zealously guarding its territories! That he rose to the level of Secretary to the Government of India in diverse

Ministries/Departments like Electronics, Renewable Energy, Small Scale and Rural Industries and Food Processing, is itself a tribute to his versatility, breadth and depth of capability and capacity born of thorough knowledge of every subject he handled, tapping into his vast network of professionals, friends, well-wishers at home and abroad. (His education at Cambridge University in the UK and Massachusetts Institute of Technology in the USA and his deep involvement in the Pugwash Conferences on Science and World Affairs in London and the South Centre in Geneva right from the very early part of his career, probably 'internationalised' him, so early!)

He was certainly not one of those born with the classic 'golden spoon in the mouth'! Yet, his pedigree was exceptional and truly 'golden'-hued: AP was the son of the legendary G. Parthasarathi, probably the greatest Guru, the country has produced in Diplomacy, Foreign Affairs, Media matters and International/ Indian/Developmental Economics & Policy, aside from his rare blend of wisdom, empathy, national strategy and priorities, at the core. (Ashok's grandfather, GP's father, was none other than the iconic Gopalaswami Ayyangar one of the nine men who wrote our nation's Constitution and who was then Nehru's right hand man on Pakistan, the erstwhile Princely States of British India, Railways and Defence.)

During some of his hectic sojourns in Bangalore, Ashok would take time out and come home for a typical (mild!) Madras/ Coimbatore-Telugu vegetarian meal. Rathna and I would hear wonder-struck, his touching recounts of the great GP's career, steeped in wisdom and holistic national service (like grandfather, like father, like son!). Our favourite, listening raptly and increasingly aghast, was AP's rendering of GP's brilliant, truthful, relationship building with both the combatants involved in the Sri Lankan imbroglio viz., the Government on the one hand and the guerrilla force called the Liberation Tigers of Tamil Elam (LTTE) seeking to carve a Tamil homeland from the present Sri Lanka in which the Tamils were a majority on the other. GP's diplomacy was so effective that he won the trust of both President Jayawardene and the supreme leader of the LTTE, Prabhakaran. However, most surprisingly, and incorrectly, GP was sidelined by Rajiv Gandhi on erroneous advice. The rest is the tragic Indo-Sri Lankan-Tamil history, still fresh in everybody's memory for the blood that flowed and the gore that followed for more than three decades, highlighted by the assassination of our youngest ever Prime Minister, Rajiv Gandhi.

New brooms often sweep clean, but history repeats itself again and again, with tragic consequences, especially when impetuous show-offs offer, nay 'impose 'miracle cures' on their respective immature and inexperienced supremos, playing with the fates of entire nations, resulting ultimately in hundreds of thousands of men, women, children and our brave trusting jawans killed, topped by 'serial' assassinations across the Palk Straits, in this most senseless chapter, of many stupid ones since (and preceding) 1947 in the continuing saga of post-Independence sub-continental history. It's so easy to forget after the twin towers collapsed, that terrorism on a mass-scale, suicide bombings and multiple assassinations actually originated in our own backyard which could have been avoided. If only Rajiv Gandhi had listened to three generations of illustrious patriots in the self-same family.

AP's early career in Delhi was dominated by his role as Science and Technology Advisor to PM Indira Gandhi, for 9 years in two spells. It was this that held AP centre-stage with all matters and all ministries connected with S&T and the nexus between science, technology, development and security policy globally and domestically. A savant, world-traveller, prolific and professional writer, an always-in-demand public speaker on serious topics, AP was comfortable and free-flowing with a wide range of 'verticals'. He has an unerring instinct to hunt out the truth from the best source, whenever he harboured the slightest doubt in his preparations for an article, a book or a lecture. Decades after my 6-years with the Nuclear Programme, starting with Vikrambhai, AP and PAG's primus inter pares, Dr Kirit Parikh (another great guru of mine) and the subsequent, wonderful and memorable 12-year stint in our Space Programme; AP would suddenly call to get the historical facts and the logic in decisions made in these, the greatest S&T programmes of the country. His book, *Technology at the Core*, describing his S&T with Indira Gandhi as her S&T Adviser is a seminal work.

However, I've discussed and argued enough with him, begging him to take the real plunge: His no-holds-barred magnum opus bursting inside him, awaiting 'pen to paper', from early childhood through his classic education with accent on Physics, right up to his post-retirement JNU saga—setting up and running his dream institution at JNU as Distinguished Professor, for Science Policy and Developmental studies. Between these extremes, these 'book ends', there'll I'm sure be revelation after revelation of how the great things that happened in and to India over a decade and more, really happened and why; how the worst that happened, really did, and why; who were the real patriots, heroes, heroines and the real traitors. How were historic opportunities (not just the sabotaged the critical and strategic National Silicon Facility) but those in War and Peace, with shades of Tolstoy, since independence in 1947 were missed, inexplicably, mysteriously. What really happened, before, during and especially in the immediate aftermath of the Liberation of Bangladesh; how exactly did India, 'unexpectedly' develop guts before December 1972, and why did a victorious, rejoicing nation, witness chickening out by Delhi, when the fully legitimate accession of Jammu & Kashmir by Maharajah Hari Singh, was finally—after nearly a quarter century of politico-bureaucratic debate and dithering—within easy grasp. With an increasingly frustrated, yet fiercely patriotic armed force in the borders, ocean and sky, awaiting just recompense for their no-warning shock, surprise and humiliation in 1962 and the fraternal enemy's transparent treachery before and since 1947, with nothing but endless debate and dithering spouting out of Delhi, an impatient and frustrated nation, wondering what the hell really was going on, each time, with the renowned politico-bureaucratic quagmire and soak-pit of Delhi! Was it the pressure from Chanakyapuri, the (Foreign) Diplomatic Enclave or directly from the by then irrelevant to Victorious India White House or (simply unbelievable!) the Kremlin, or whatever else on Earth, for God's sake!?

The nation awaits answers with absolutely none forthcoming to this day. What exactly happened in Tashkent after victory? No answer! Who were the truly great Defence & Foreign Ministers and their Secretaries, India has ever produced, and who were the mediocrities or worse? No answer. After 1962, at least six post-Independence generations (starting with those alive at that time) have miserably failed India. They were also led to failure by their elected 'leaders' and their cronies. Personally, this is my only hope left for the truth to come out, so that I can tell the truth to my own sons, grandchildren and numerous students, and if possible and relevant, also my own little part, in strengthening my motherland. More than that, the great part played by beloved AP for six and a half decades (and counting) of dedicated, totally result-oriented, continuous national service. I know AP enough to implicitly trust him, totally, that one day he will edify several puzzled generations of Indians, who first yearn for the unvarnished truth (including what exactly happened to Netaji!). AP was certainly privy to many, if not most, of the shenanigans in the dark, opaque, non-transparent, power centres of Delhi, who believe in and swear by 'terminological in exactitudes' galore, all their lives! My hunger for AP's magnum opus is therefore, total and I catch myself praying that before one kicks the bucket beloved AP will 'spill the beans'!

AP's stewardship of the unique Programme Aimed at Technological Self Reliance (PATSER) in the Department of Scientific and Industrial Research (DSIR), Ministry of Science and Technology, when he was the Additional and later Special Secretary, gave me an opportunity to request him to open the programme to the private sector. He required little persuasion, and so it was that DSIR dispersed through PATSER ₹20 crore a year, as far as 1995. What is more PATSER was the very first to start financing private companies with up to as much as 40 per cent eligible from provision as grant. Meanwhile, my own former client in the Technology Development and Information Corporation of India (TDICI) also led hi-tech small and medium sized companies coming to PATSER (DSIR) for such unique grant support, with which I kick-started the Venture Capital Movement in the country in 1988. M/s Turbotech's Low-Cost Gas Turbine project was supported by PATSER, leading to a long tripartite relationship between Krishna Kumar, the brilliant engineer from Garret Aerospace, AP and me on the board!

Iconic however was AP's brilliant stewardship as Chairman of Central Electronics Ltd (CEL) and the National Research Development Corporation (NRDC) [I had preceded him by a year or so as director in the latter, asked by Professor Dhawan to represent the Space Programme in it]. When my senior in IIT (Madras) N.K. Sharma was selected by AP to be the Managing Director of NRDC, AP became the part time Chairman of the Board of Directors of NRDC, we wanted none other than him to be the chairman because of the unique character—it had licenced technologies to industry both public sector and private sectors; to large, medium and small scale and start-ups. A glorious innings of NRDC, never before or after, ensued. Dr Gopalan (IDBI) and I were the founder-directors of BCIL, the Biotech Company to support Biotech in the country, in 1990. My Delhi trips then became more frequent, with both the regular BCIL & NRDC Board meetings, enabling me to develop greater and deeper professional understanding of AP's missions in life.

During my innings of 18 years in Atomic Energy and Space, resigning only to kick-start the Venture Capital, Angel Investment and Incubation Movement, a long sought dream in my mission of entrepreneurship development in the country, AP's tacit and active support was always there. What's more after my stint successfully starting and running the nation's very first venture capital funding company, AP himself suggested that I should think of succeeding N.K. Sharma as MD of NRDC. This was a great option for me at that time, though this and a similar option in BCIL, were not pursued in part due to my family compulsions in Bangalore. Santy and Kirit Parikh were my other Gurus, who were ready, years earlier, with options for me in great institutions in Mumbai and elsewhere. That these great, noble patriots retained my profile always in their minds just as AP did, was a multiple blessing! I'm always indebted to them.

In the Ministry of Non-Conventional Energy, a subject very close to AP's heart, even today, among the myriad initiatives he took, notable was the reform of the Indian Renewable Energy Development Agency (IREDA), the funding arm of the Ministry and Chaired by AP. I introduced to AP the Hyderabad based Company Renewable Energy Systems (RES), active in solar photovoltaic energy modules and systems, solar thermal systems and Nickel Cadmium (NiCad) and Thermal batteries development. A lifelong professional relationship between RES and AP followed, including AP asking them to set up an indigenous technology based NiCad plant for Punjab Batteries Ltd. in Mohali.

On a purely personal note, never written before, I was an angry boy from early childhood hearing about the sufferings of my Nellore Uncles, 8 of them, and their youngest brother, my father, during the Freedom Struggle and the World War II. They were tortured in 'British' jails by Indian jailors, then, suddenly at revered Gandhiji's calling, and the orders of their eldest sibling, some of them crossed the Kala Pani to fight on the side of their former jailors! Sadly, they never returned. The remaining brothers had to run the family business, the eldest, their undisputed leader, suffering from the effects of torture in incarceration. My hot-tempered father, the youngest, also joined the British-Indian army, but instead of going to the front was forced to go underground, having slapped his British officer for an insult to Indians!

For me, the humiliation at the hands of the Chinese in the War of 1962 when I was doing my Pre-University Course in Loyola College, in Chennai rankles to this day! From this nationalist origin, and the best education the country could provide, I happily and proudly served in the Nuclear and Space Programmes, shunning the private sector, determined that one day I should definitely be involved in assisting my country not only to be self-reliant, but powerful, never again to be subjugated by any external force, making myself a devoted servant of Indian S&T, as the tool for real success of my beloved nation. In my lifelong journey with this ideal, I am ever confident that AP and my other Gurus and countless friendships earned, in all generations, will bless my efforts and join in any success I may garner! Let the failures, very few so far, be my own! God bless Ashok Parthasarathi! And like Abu Ben Adem, may his tribe increase!